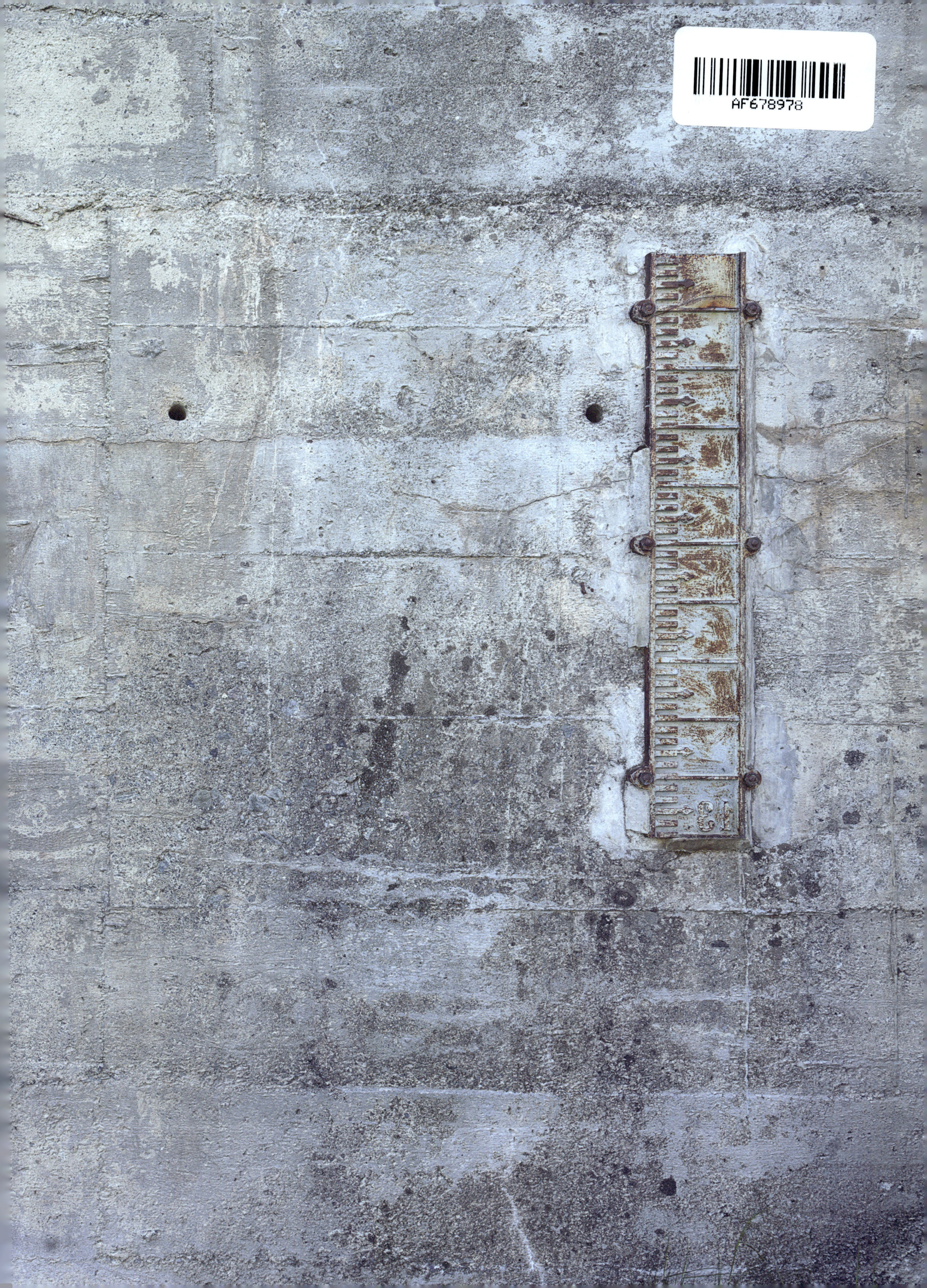

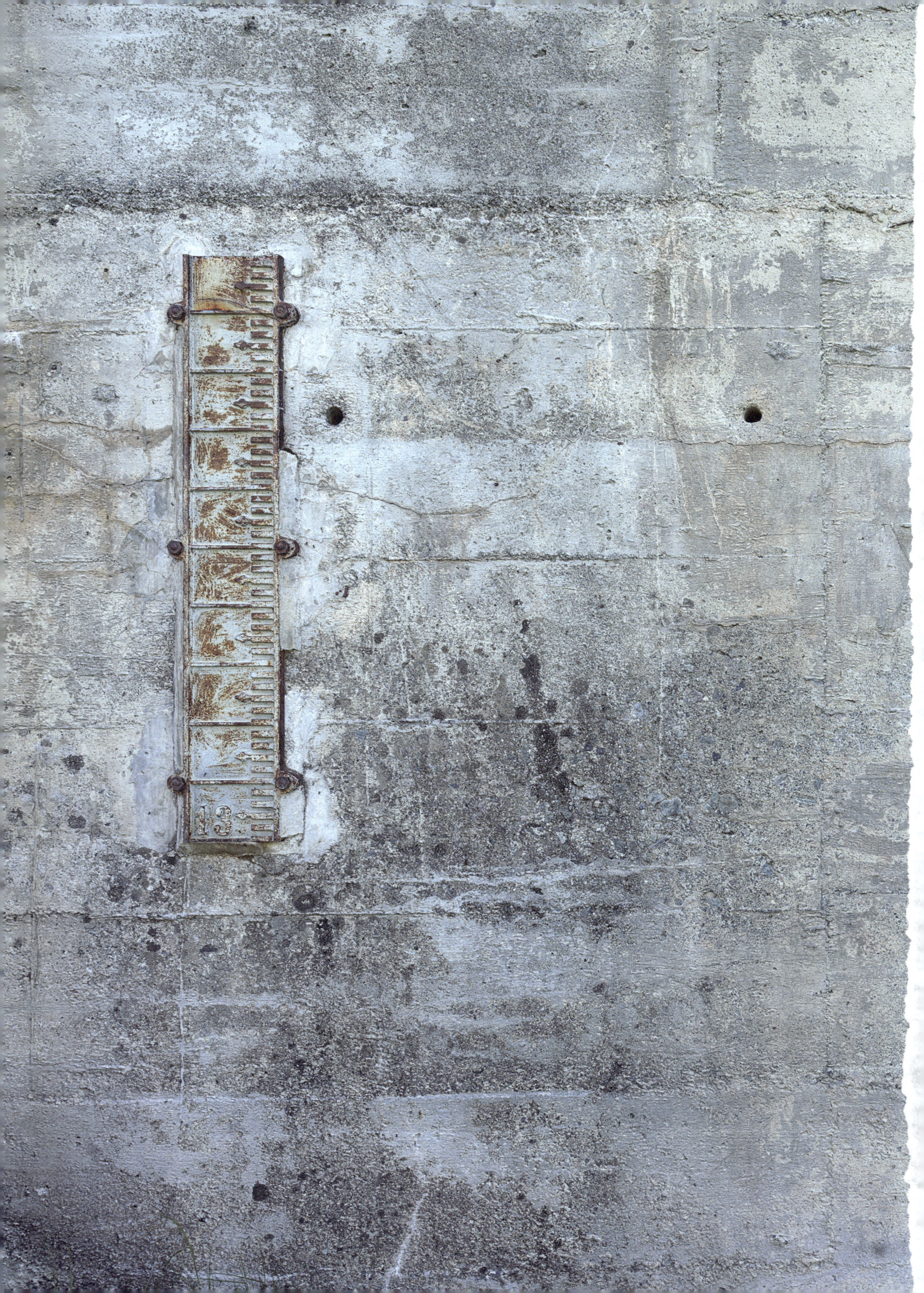
13

Simon Walther

StauWerke

Monuments of Power

Eine Hommage an die grossen Mauern
Mit einem Vorwort und Essay von Köbi Gantenbein

An homage to the great walls
With a preface and an essay by Köbi Gantenbein

Herzlichen Dank für die Unterstützung.
Thank you very much for your support.

Schweizerischer Wasserwirtschaftsverband
Association suisse pour l'aménagement des eaux
Associazione svizzera di economia delle acque

ERNST GÖHNER STIFTUNG

1. Auflage 2020 | 1st edition 2020
Simon Walther, Wattwil und Maloja

ISBN 978-3-7165-1849-6

Fotografie | photography: Simon Walther, simonwalther.ch
Text | text: Köbi Gantenbein, hochparterre.ch
Gestaltung | layout: Simon Walther; Pascal Walther, 2plus.ch
Übersetzung | translation: Cosima Talhouni
Korrektorat | proofreading: Sophie Steybe

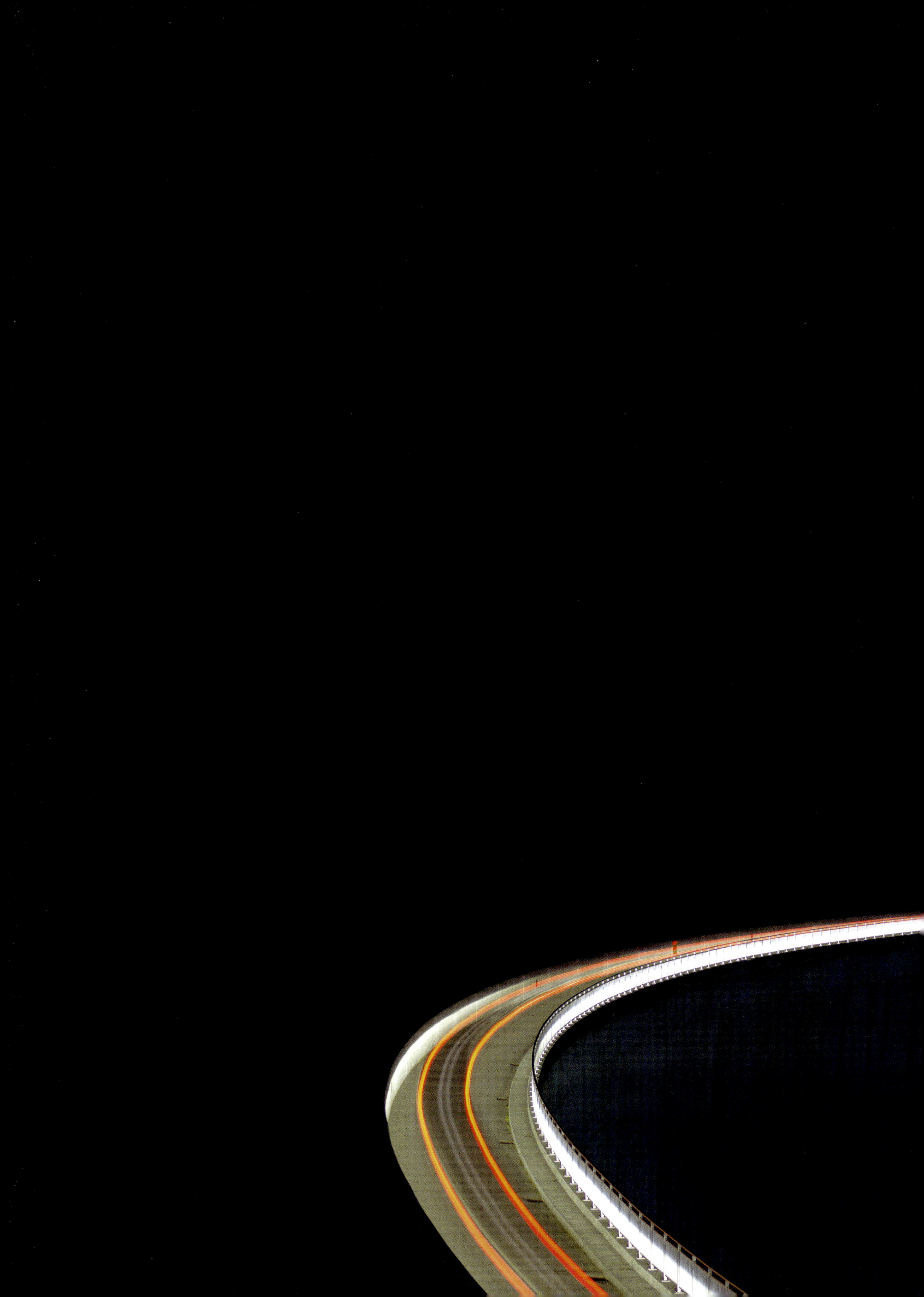

Der Mythos des grossen Bogens

Die politische und wirtschaftliche Verknüpfung von Unterland und Berggebiet hat keine eindrücklichere, mächtigere und komplexere Architektur als das Wasser-Kraftwerk. Hoch oben in den Bergen die Staumauer, die als Sperre mit tausenden Tonnen Beton in einem Bogen am Ausgang des Tales steht – über die Jahre immer schöner patinierter Kunststein, verwachsend mit der Landschaft. Weiter oben im Gebirge und in Nebentälern Fassungen, Auffangbecken und viele Kilometer Stollen, durch die das versammelte Wasser in den grossen Stausee fliesst. Von der Mauer aus wieder Stollen, talauswärts und bergdurchbohrend hin zu Turbinenstationen. Eine um die andere ist dem Gefälle nach gebaut und viele Kilometer nach dem Stausee verlässt das Wasser in wild schäumenden Bogen die letzte Turbine und macht sich unterwegs zum Meer. Und in die Stadt, die Büros, die Wohnungen, die Fabriken, die Banken, die Computer fliesst endlos der elektrische Strom.

Die Gewichtsstaumauer Pérolles an der Saane bei der Stadt Fribourg läutete 1872 ein Kapitel Schweizer Geschichte ein – dort steht die erste Staumauer in Europa, gebaut in Beton. Fünfzig Jahre später wieder ein Markstein im Kanton Freiburg. Am Lac de Montsalvens in der Nähe des Jaunpasses konstruierten die Ingenieure die erste Bogenstaumauer Europas mit horizontaler und vertikaler Krümmung. Dann war kein Halten mehr – heute gibt es in der Schweiz mehr als 200 Talsperren. Sie sind Teil eines komplexen Netzes aus Fassungen, Ausgleichsbecken, Stollen, Turbinenhäusern, Pumpanlagen und Stromleitungen. Die Wasserkraftwerke produzieren in 650 Zentralen 36,3 Terrawattstunden Strom – kein Mensch kann sich diese Zahl anschaulich machen – sie ist gross, stark und viel.

Fotografie hilft uns, Bilder im Kopf zu konstruieren, damit wir uns über das Unvorstellbare eine Vorstellung machen können – natürlich im Sinn des Fotografen, der das Bild bestimmt, den Apparat aufstellt und im richtigen Moment abdrückt. Simon Walther kann das vorzüglich. Denn seine Fotografien geben dem komplexen System Wasserkraft eine Anschauung. Er konzentriert sich auf die Orte hoch oben im Gebirge, wo die Energie, mit der wir weit unten das Licht anzünden, aus Bergbächen versammelt und aufgestaut wird. Er untersucht und erforscht vorab die erhabenen Talsperren, verzaubert von Nebel und Winterstimmung. Diese Bilder erklären den Respekt des Fotografen vor der grossen technischen Leistung der Ingenieure und Arbeiter und sie überhöhen ihr Werk – sie sehend, denke ich: «Diese Mauern waren immer schon da.» Wie der Fels, der Baum, der Bergbach.

So funktionieren Mythen, sie erhöhen von Menschen Gemachtes in Ewiges. Und die Stauseen gehören denn auch zweifellos zum Bestand der Schweizer Mythen wie das Matterhorn, der Gotthardtunnel für die Eisenbahn oder die AHV. Doch Simon Walther nimmt uns an die Hand und führt uns nahe zur Sache. In ungewohnten Bildausschnitten und -perspektiven sehen wir hinein in das Menschenwerk; sehen, wie die Kalkulation des Bauingenieurs die Bogenmauer innen abtreppen lässt, wie der Beton altert, wie Messapparate installiert und gewartet werden müssen und wie das Postauto über die Mauerkrone fährt. Doch auch an und hinter der Mauer ist Simon Walther dem Mythos auf der Spur – die Zauberbilder übers Wasser sind nur möglich, weil die Mauer es staut. Die eindrücklichen Aufnahmen der halb leeren Stauseen sind keine Wunden in der Landschaft mehr, sondern das Kraftwerk hat neue, faszinierende Landschaften geschaffen, die hervorkommen, wenn die Energielast ins Tal gedonnert ist und die verschwinden, wenn sie wieder aufgeladen wird. Und darum auch hier – die gewaltige Installationen, die Arbeiter und Baumaschinen in archaische Täler gestemmt, gesprengt, gebaggert und betoniert haben, sind eins geworden mit der Natur.

So fügt Simon Walther die Kraftwerklandschaften ein in die Kultur- und Zeitgeschichte der Schweiz, die mit der Wasserkraft eng verbunden ist. Als produktive Kraft, die zu 57 Prozent aus dem Wasser kommt und die Wirtschaft und Gesellschaft im Mittelland vom Berggebiet abhängig macht, denn die Hälfte des Wasserstroms kommt aus Graubünden, Tessin, Uri, Wallis, Bern und Freiburg. Mein Essay auf Seite 34 für dieses Buch ist denn auch aus der Perspektive eines geschrieben, der zur Zeit des grossen Kraftwerkbaus im Berggebiet aufgewachsen ist und heute dort lebt. Er schildert den politischen und wirtschaftlichen Skandal, dass die Abgeltung für die Wasserkraft ans Berggebiet keineswegs genügt; der Essay taucht aber wie Simon Walthers Bilder auch ein in die Faszination dieser Bauten und ihrer Geschichte, die einige Täler und Landschaften in den Alpen auf immer und ewig verändert haben, die aber der Gesellschaft und den Leuten in den Bergen auch allerhand gebracht haben und bringen.

Köbi Gantenbein

The Myth of the Great Arch

In the political and economic interaction of lowlands and mountainous regions there is no architecture more impressive, powerful and complex than a hydroelectric power station. High up in the mountains, the arched retaining wall stands as a dam made of several thousand tons of concrete at the valley mouth – artificial stone that grows in beauty and patina with the years as it becomes immersed in the landscape. Further up in the mountains and in secondary valleys, there are embankments, retention basins and many kilometers of ducts through which the collected water flows into the large reservoir lake. From the wall also, many ducts extend out of the valley and penetrate mountains to reach water turbine stations. Each is built according to the downward slope, and many kilometers beyond the reservoir lake, the water violently exits the final turbine in a wild and foamy curve and heads to the sea. At the same time, the generated electric power flows endlessly into the city, the office, the homes, the factories, and the computers.

The gravity dam Pérolles on the Saane/Sarine near the city of Fribourg initiated an era of Swiss history in 1872 with the construction of Europe's first retaining wall made of concrete. Fifty years later, another milestone was constructed in the canton of Fribourg. On Lac de Montsalvens near the Jaun Pass engineers built Europe's first arch dam with a horizontal and a vertical arch. After that there was no stopping the Swiss – today there are more than 200 dams across the country. They are part of a complex network of embankments, regulating reservoirs, ducts, turbine houses, pump systems and power lines. At 650 control centers the hydroelectric power stations produce a total of 36.3 terawatt hours of electricity – no one can visualize this number – it is huge, powerful, and plenty.

Photography helps us construct images in our heads to allow us to imagine the unimaginable – of course as intended by the photographer who determines the image, sets up the camera and pushes the button in the right moment. Simon Walther excels at this. His photographs visualize the complex system of hydropower plants. He focuses on the locations high up in the mountains where the energy that powers our lights far below is collected from mountain brooks and retained. First he examines and explores the lofty dams with their enchanted fogs and winter moods. These pictures express the photographer's respect of the great technical achievement of the engineers and workers and elevate their work – looking at them makes me think: "These walls have always been there." Just like the rock, the tree, and the mountain brook.

This is what myths do. They elevate man-made things into something eternal. Therefore the reservoir lakes are undoubtedly also part of the stock of Swiss myths like the Matterhorn, the Gotthard Tunnel for the railway or the old age and survivors insurance. However, Simon Walther takes us by the hand and leads us next to the constructs. Unusual image details and perspectives allow us to look inside the human-made work, to see how the calculations of the constructional engineer allows the arched wall to contain benched steps on the inside, how the concrete ages, how measuring instruments have to be installed and maintained and how the mailman drives over the crown of the wall. At the same time, Simon Walther also examines the myth on and behind the wall – the magical pictures of the water are only made possible by the wall that retains it. The impressive photographs of half-empty reservoir lakes are no longer wounds in the landscape, instead the power plant has created new and fascinating landscapes that emerge when the collected energy has thundered into the valley and that disappear as it is being collected again. This is why here also, the monumental installations, which workers and construction machines have heaved into archaic valleys, detonated, excavated, and cast in concrete have become one with nature.

Thus Simon Walther integrates the hydraulic power plants into the cultural and contemporary history of Switzerland that is closely associated with hydraulic power. A productive power, 57 percent of which originates from water, that makes the economy and society of the midlands reliant on the mountainous regions, because half of our hydraulic power comes from the cantons of Grisons, Ticino, Uri, Valais, Bern and Fribourg. My essay on page 38 of this book is also written from the point of view of someone who grew up in the mountains during the height of power plant construction and who lives there today. It describes the political and economic scandal that the compensation for the mountainous region for using its water is far from sufficient. However, similar to Simon Walther's photographs, the essay also delves into the fascinating aspects of these constructs and their history that has changed some of the Alps' valleys and landscapes for all eternity, yet has brought and continues to bring a variety of benefits to the mountains and their residents.

Köbi Gantenbein

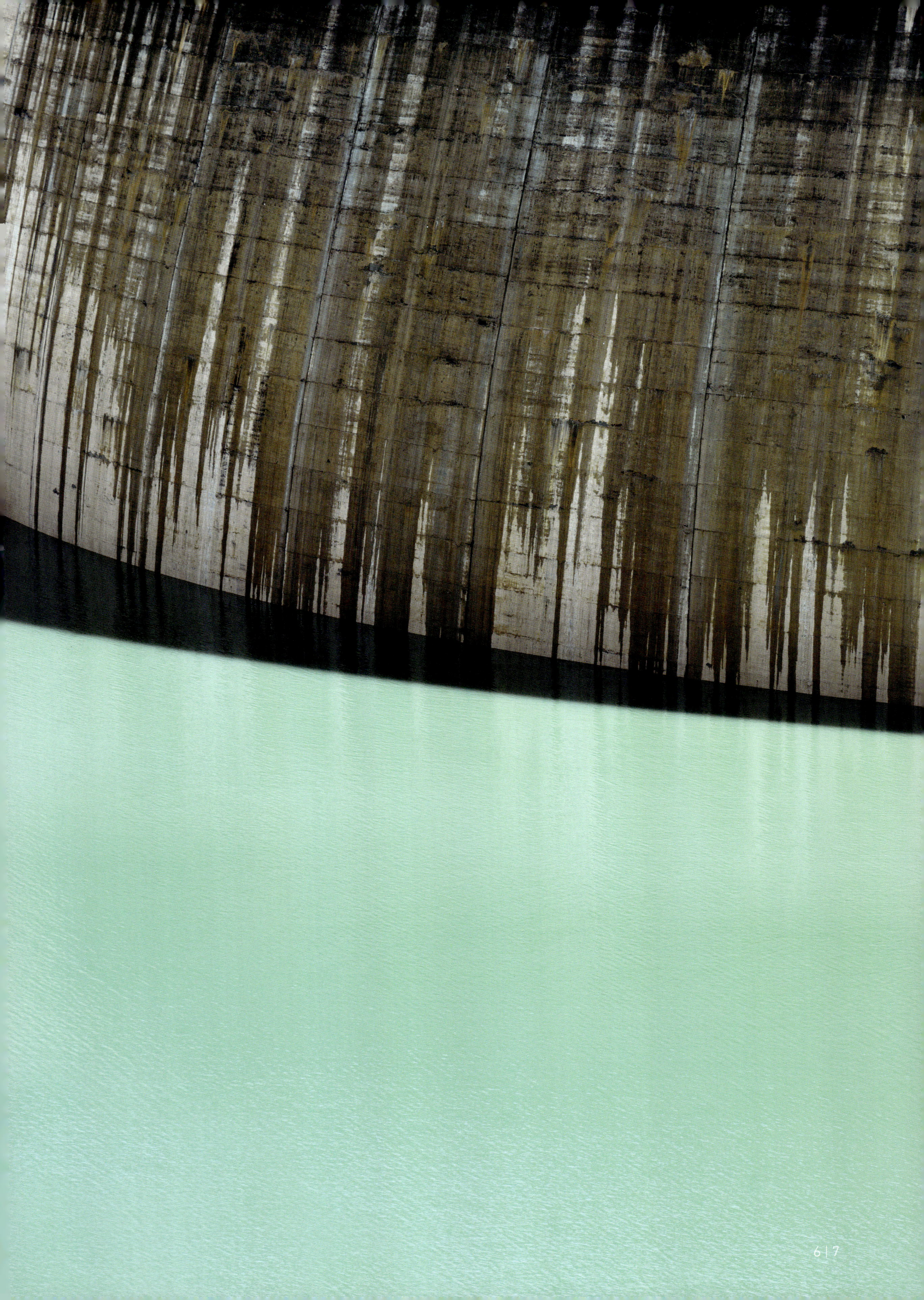

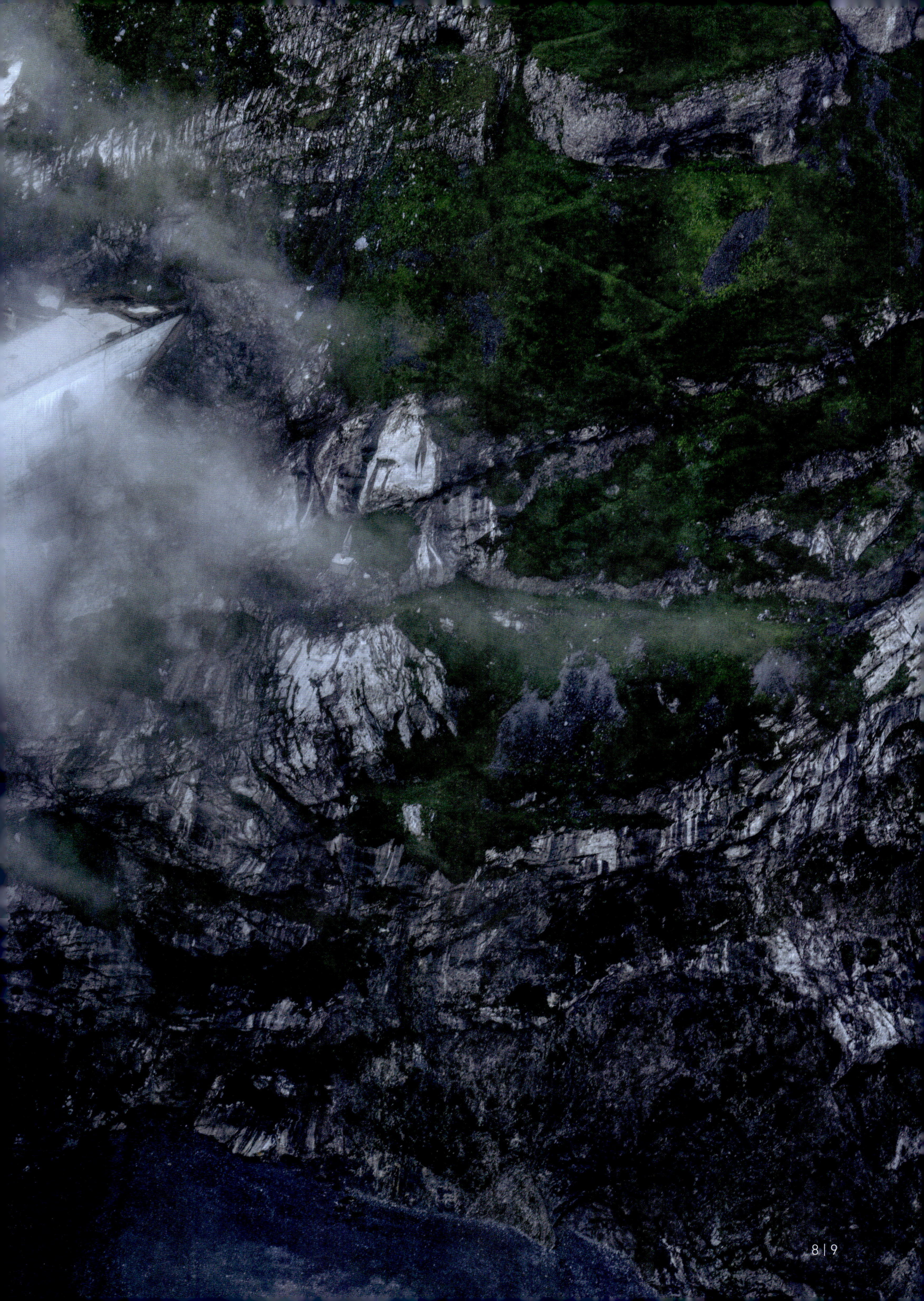

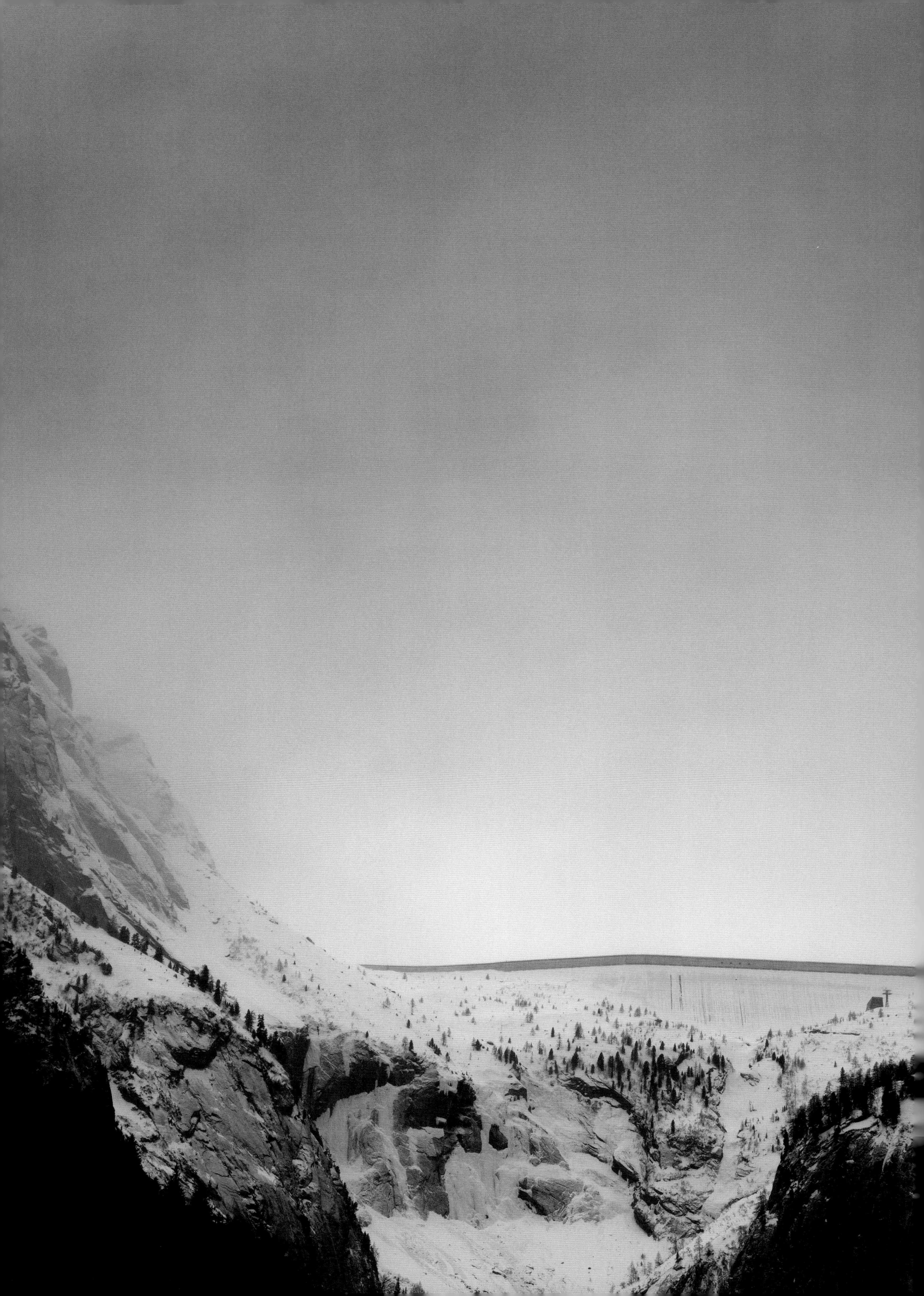

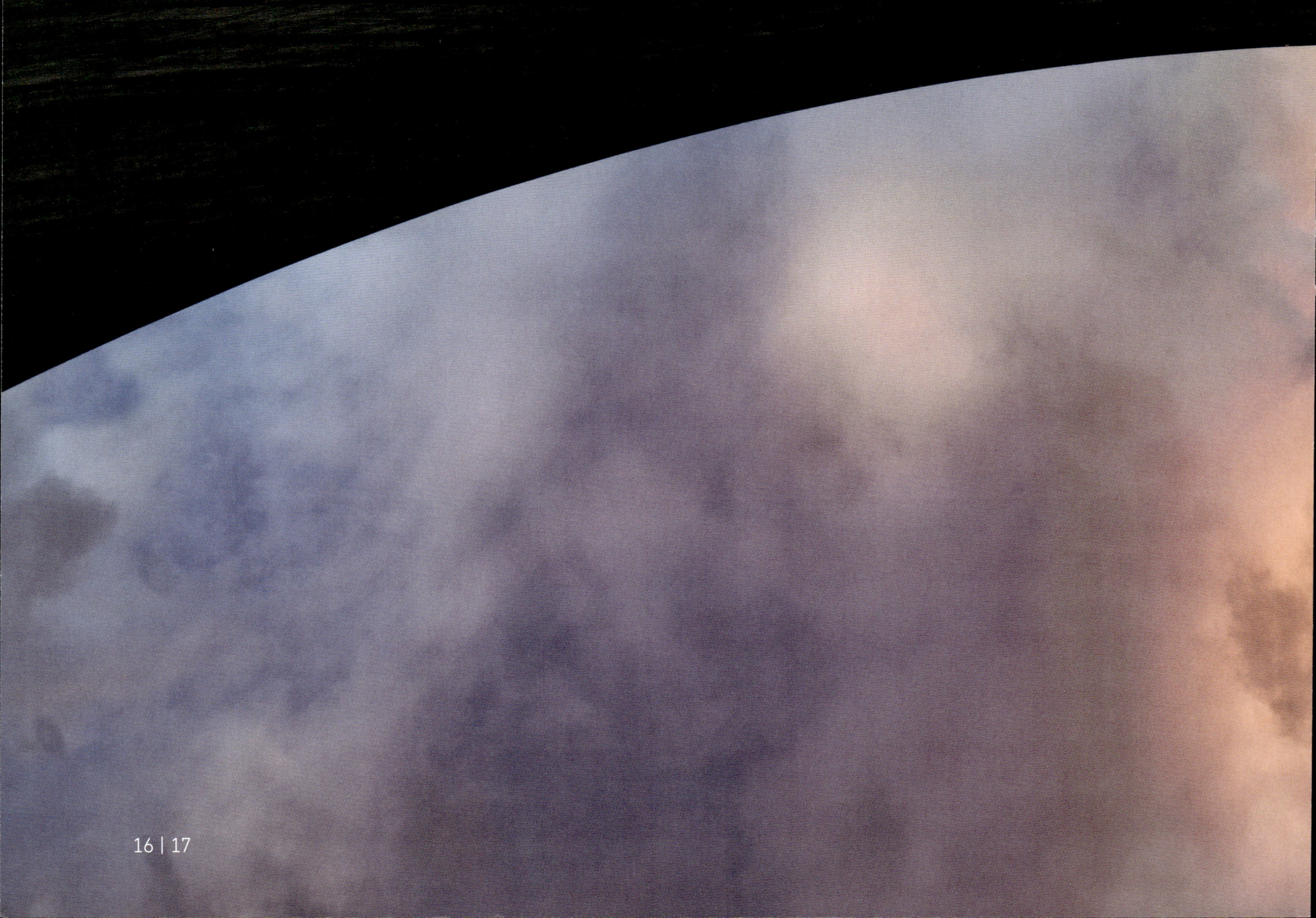

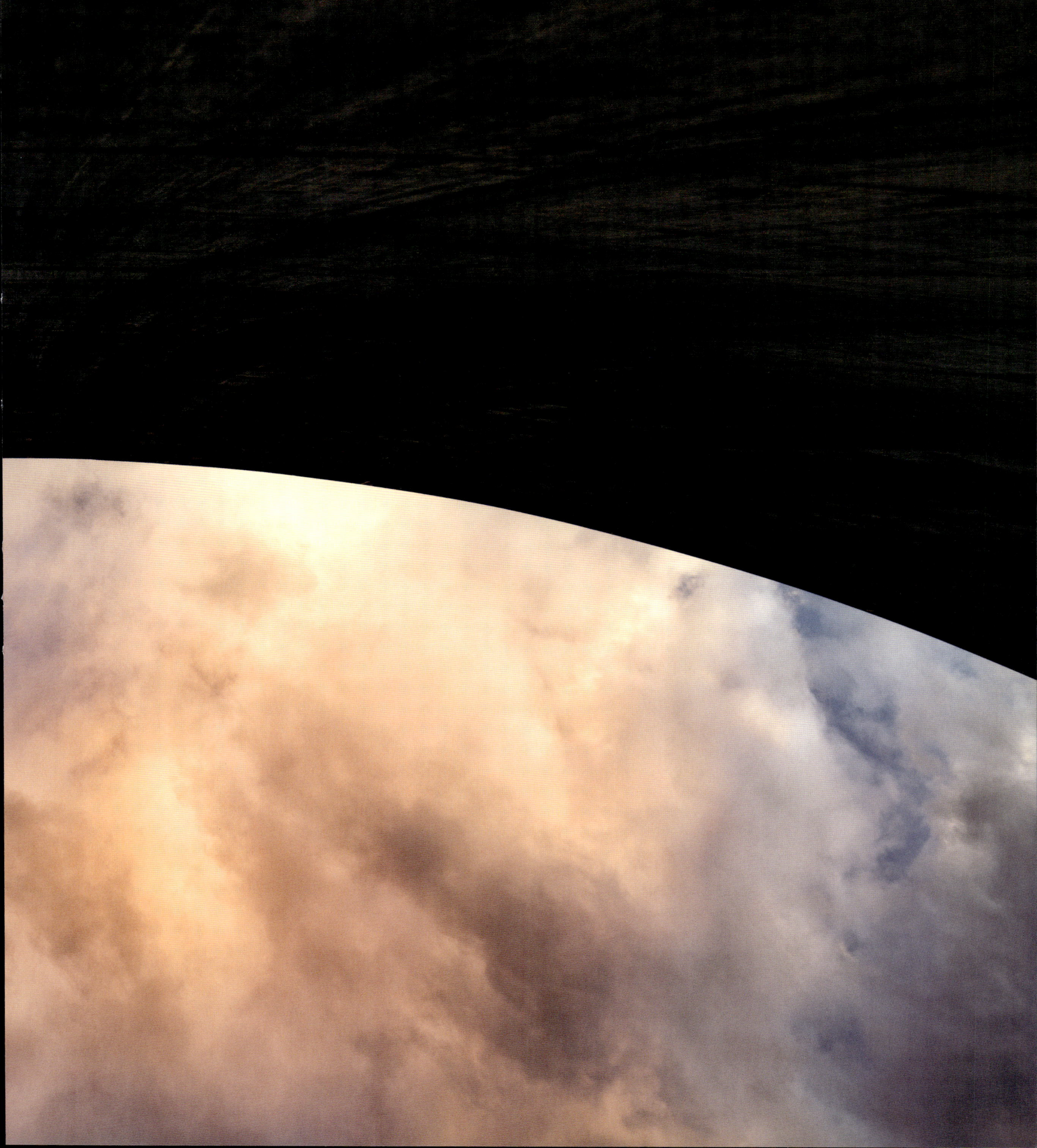

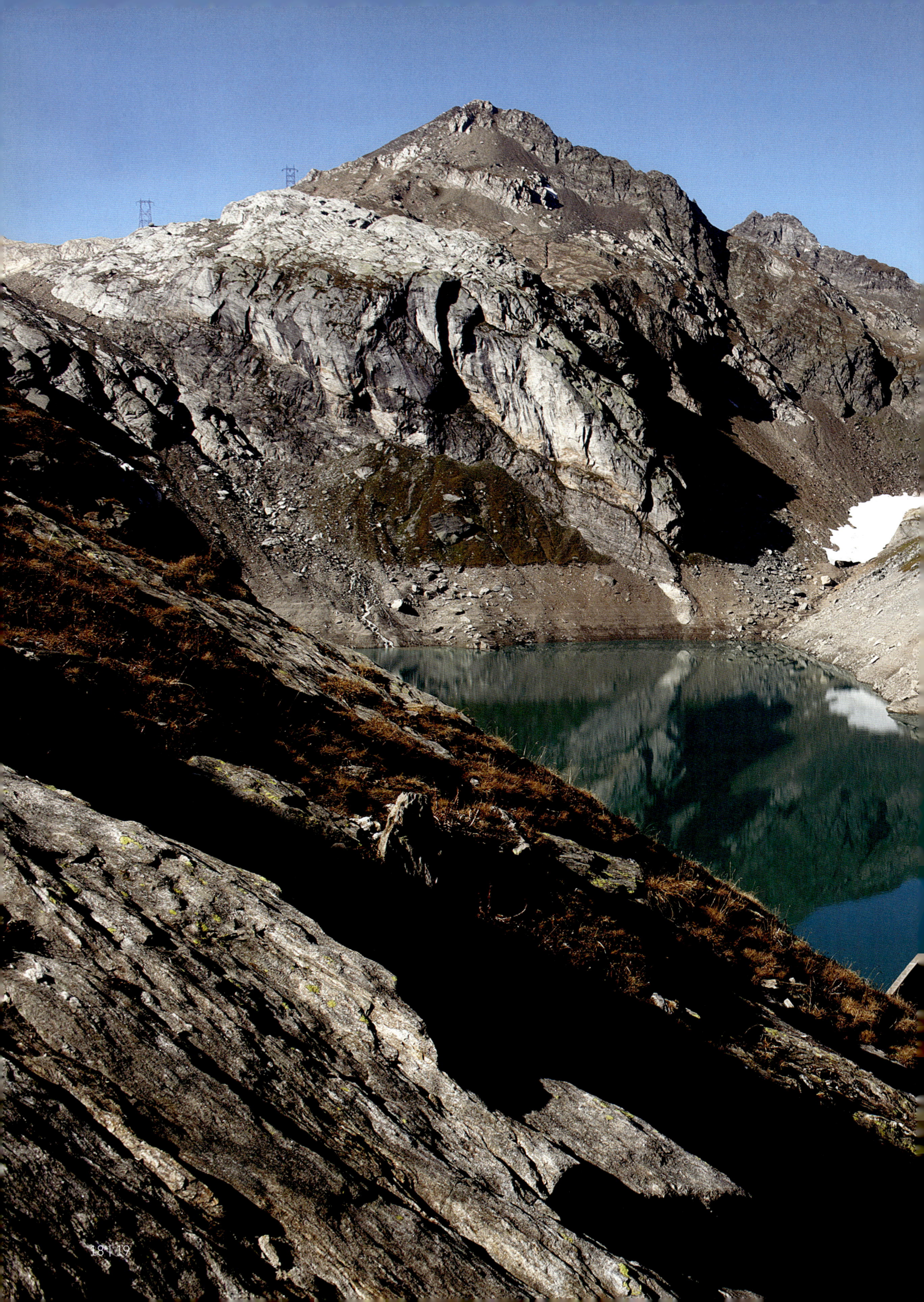

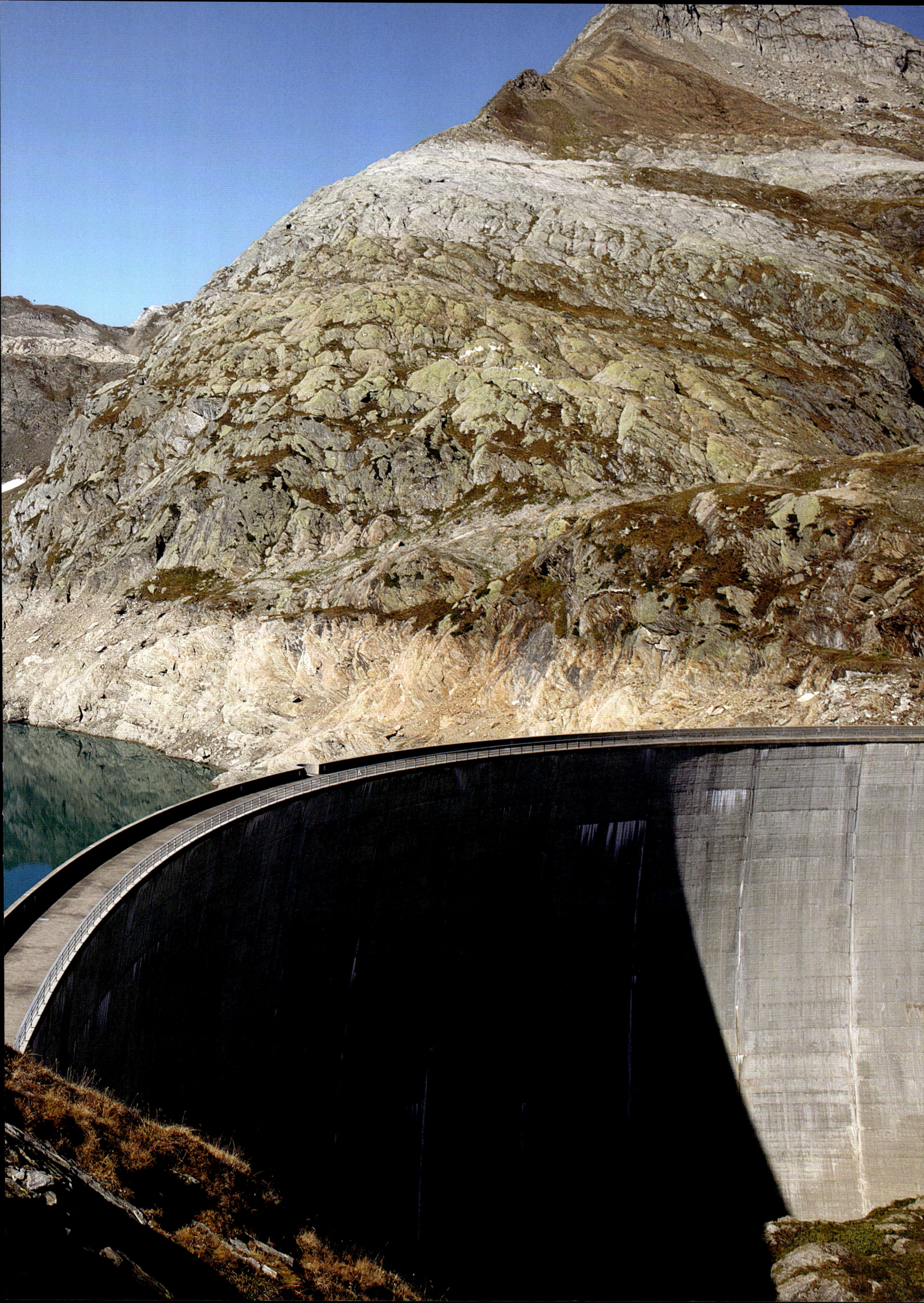

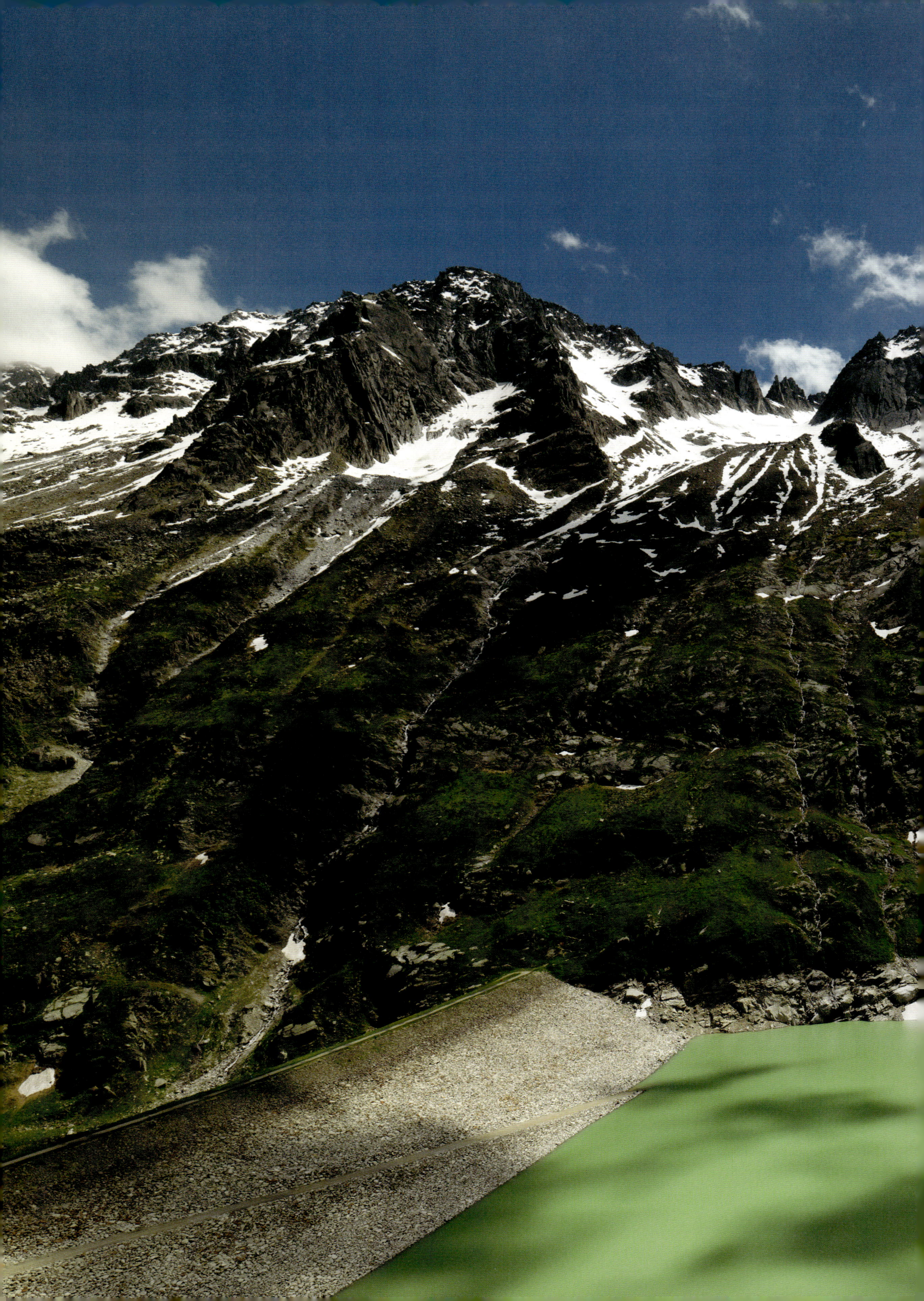

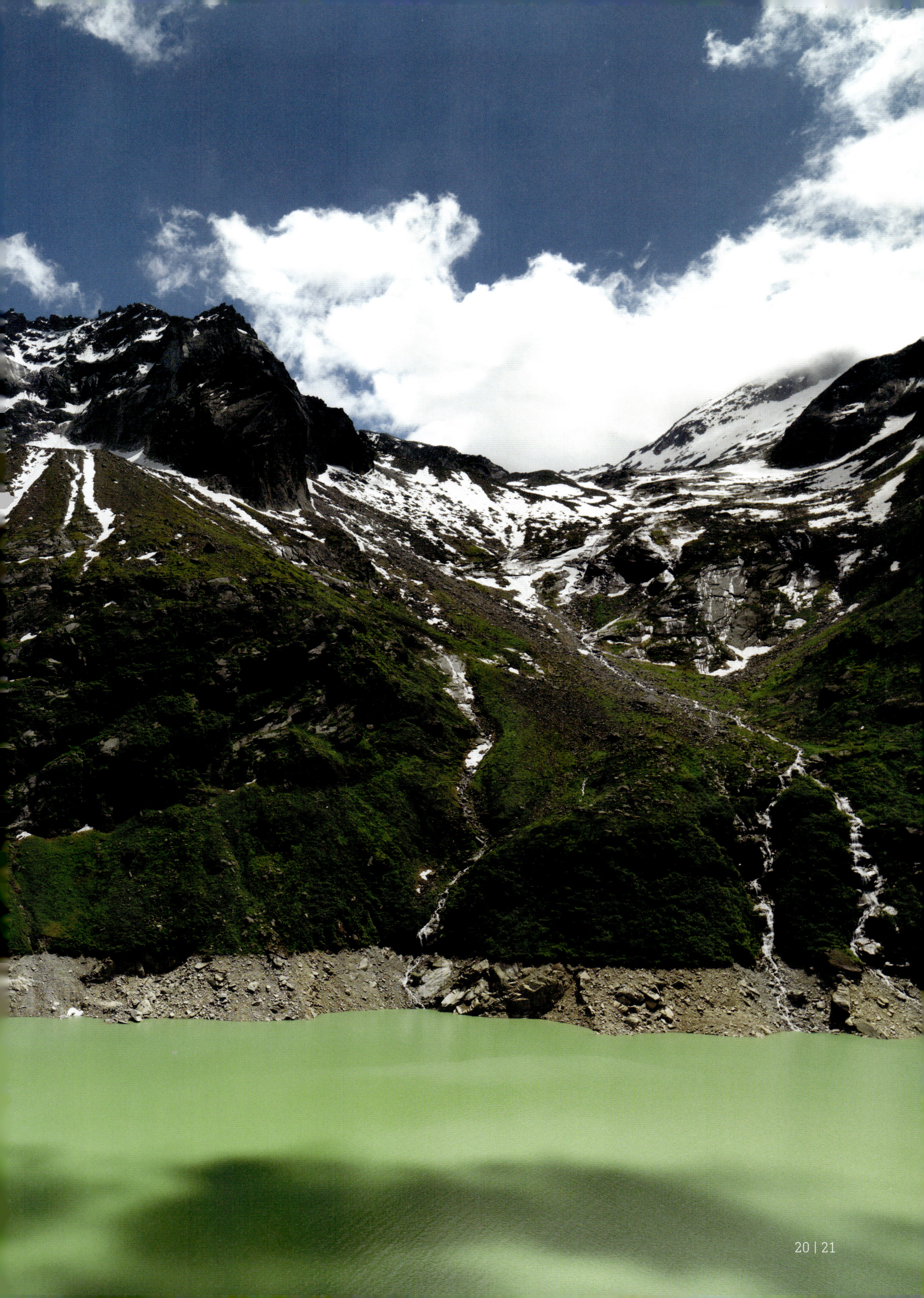

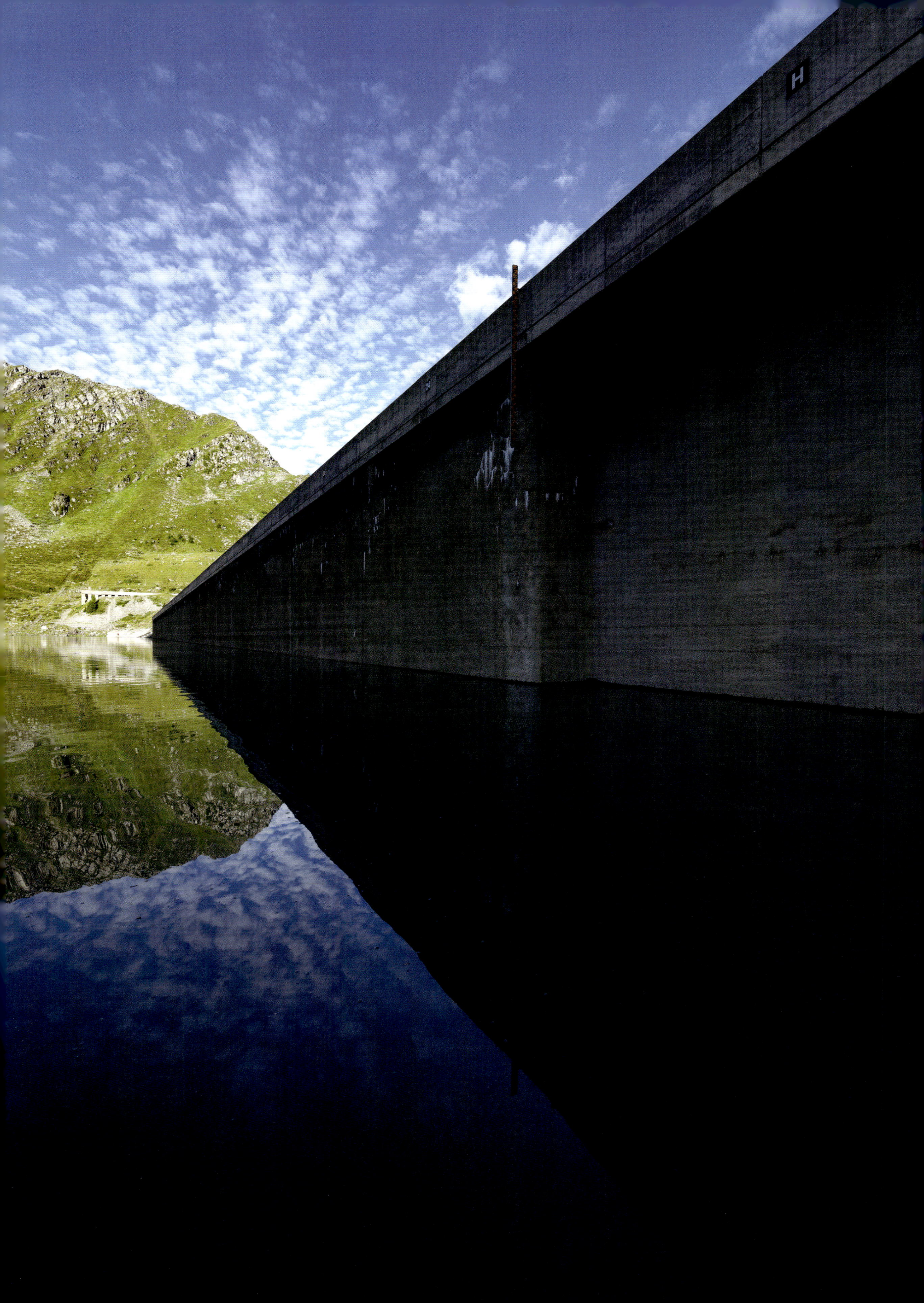
H

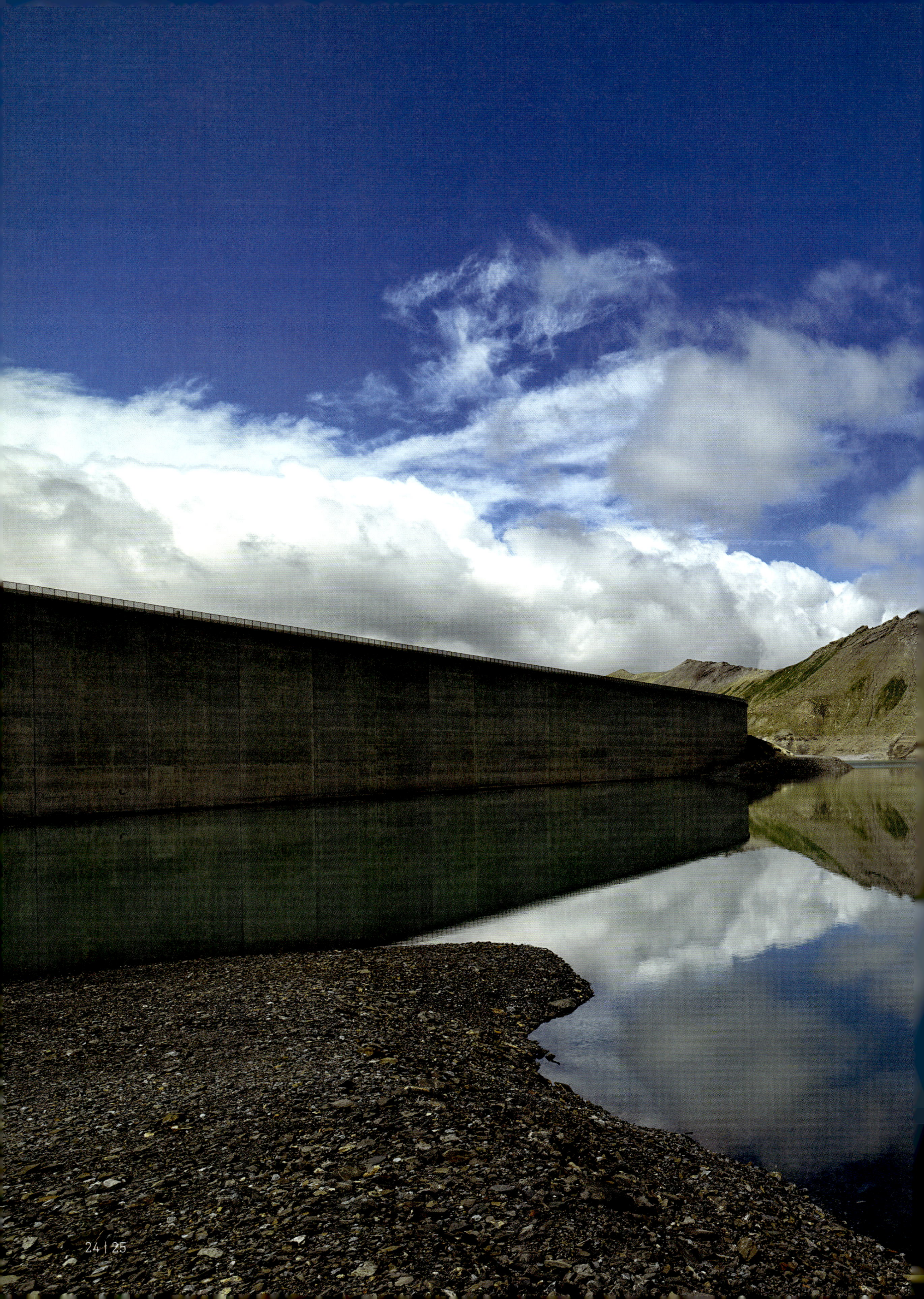

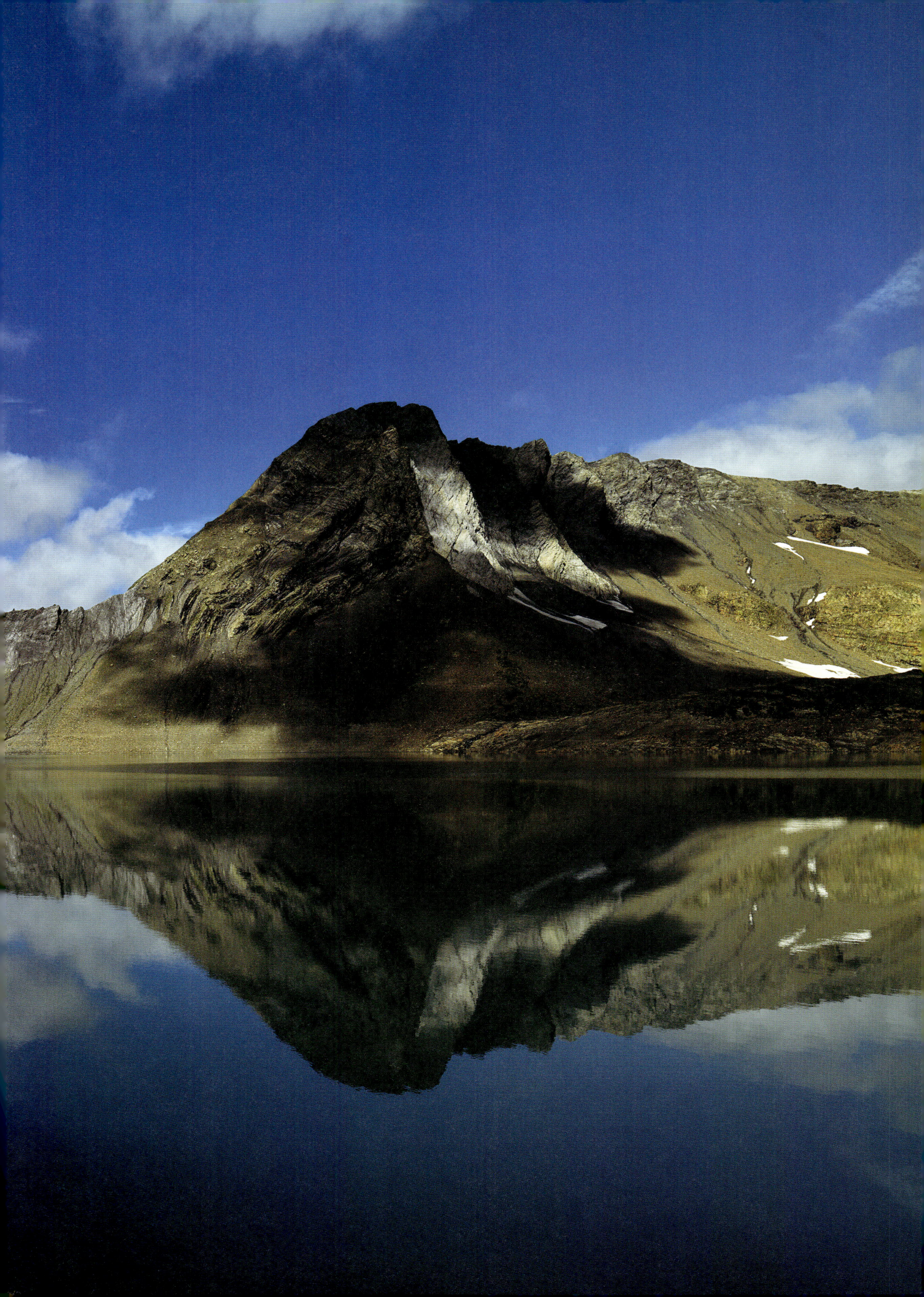

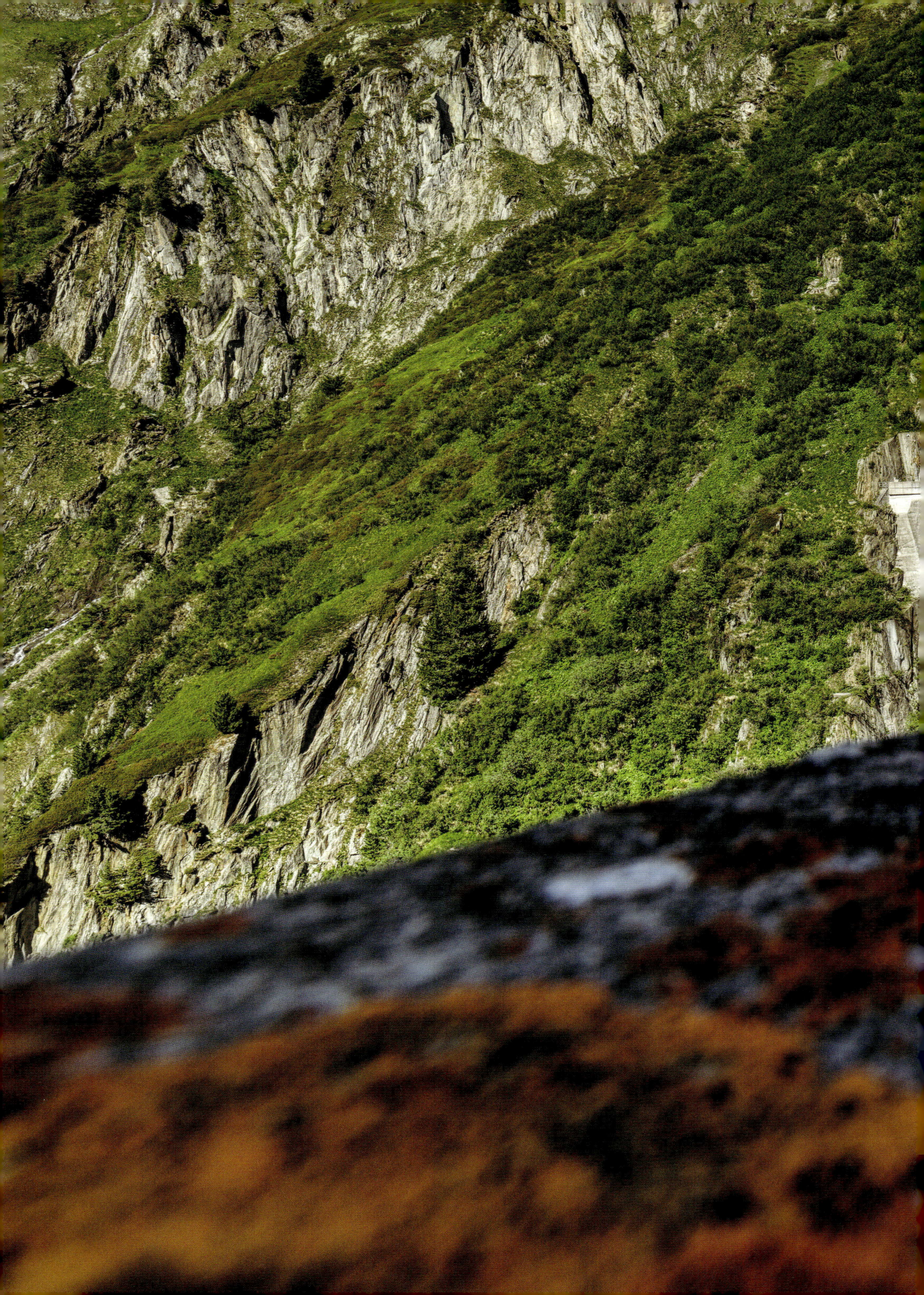

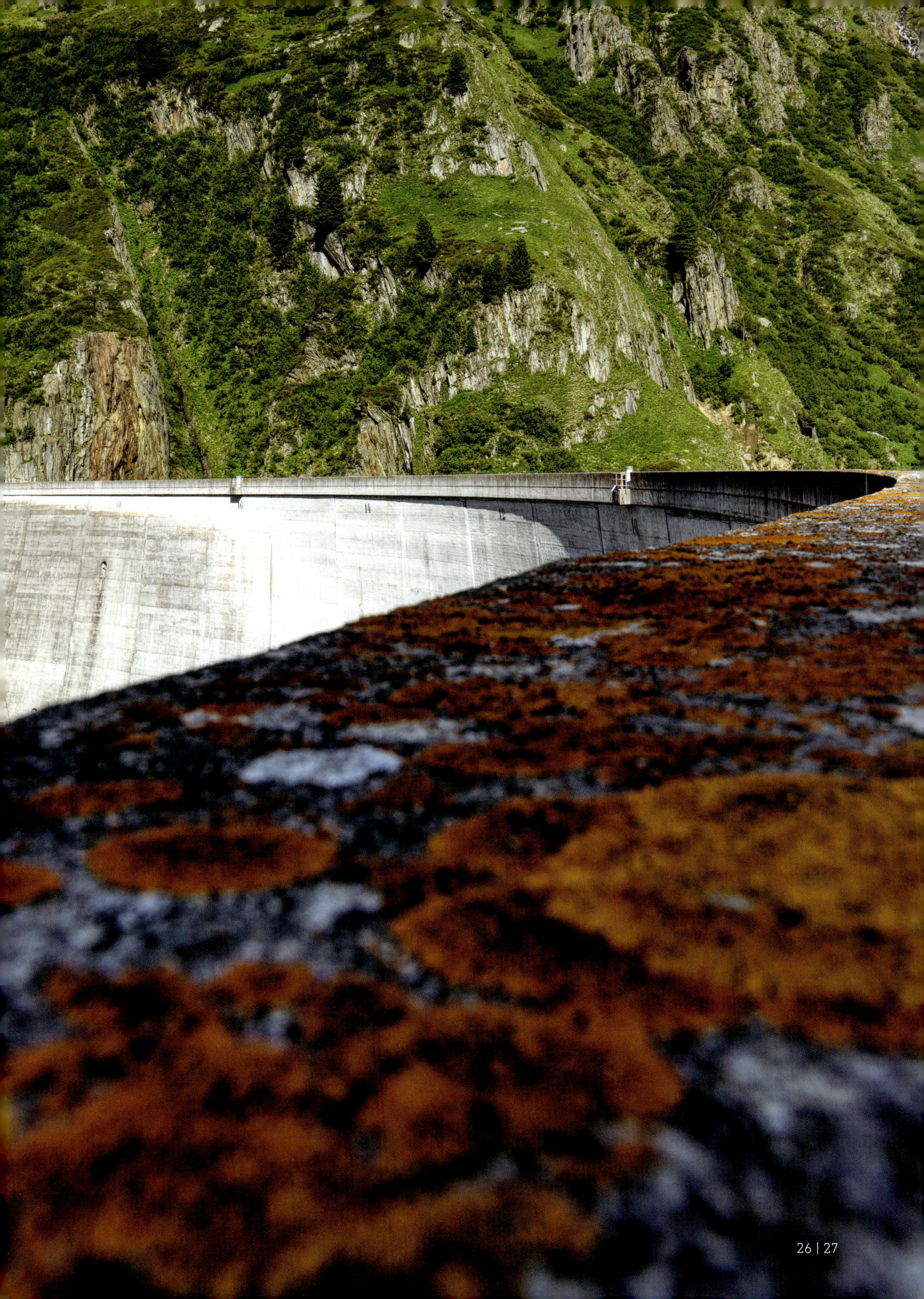

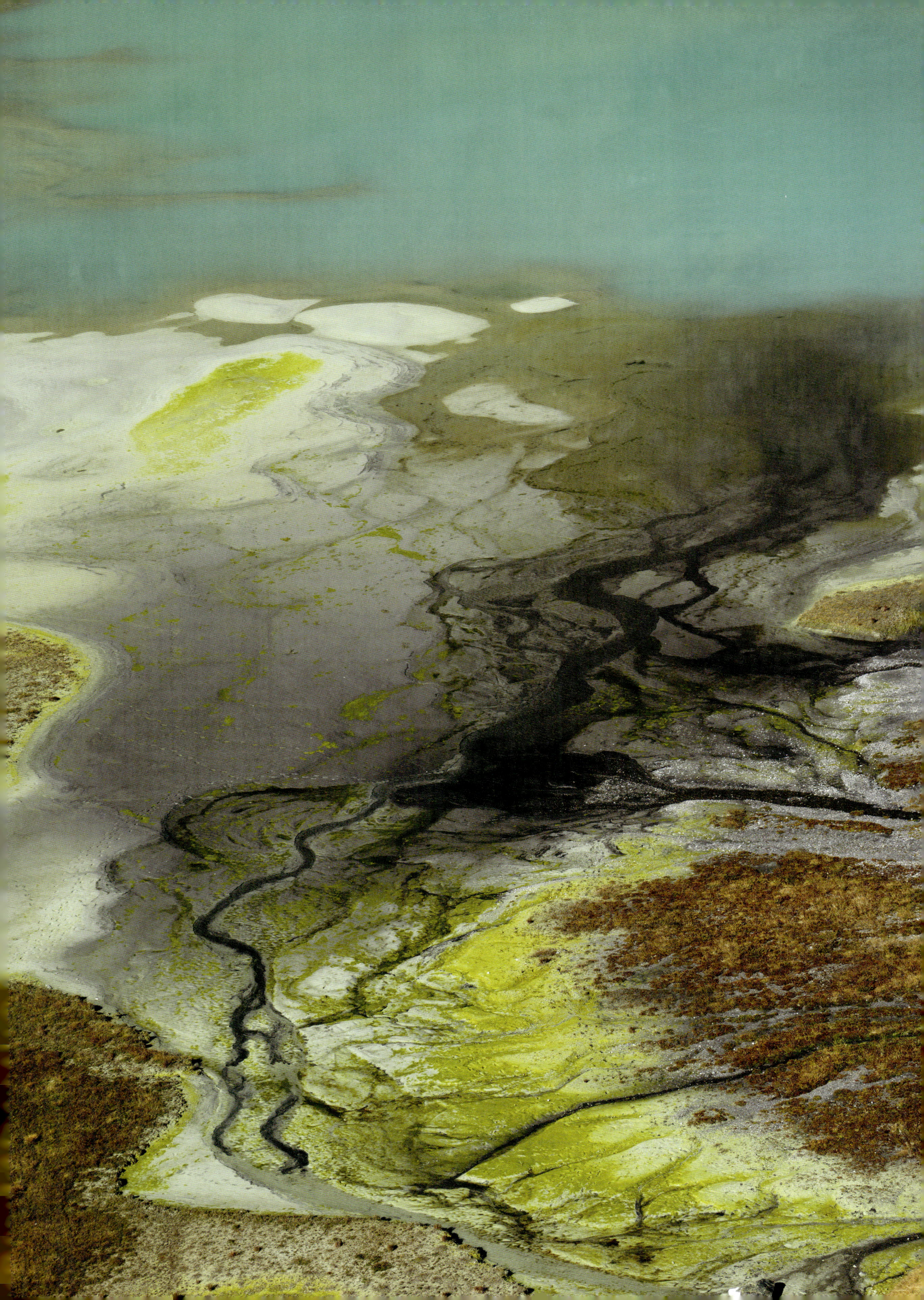

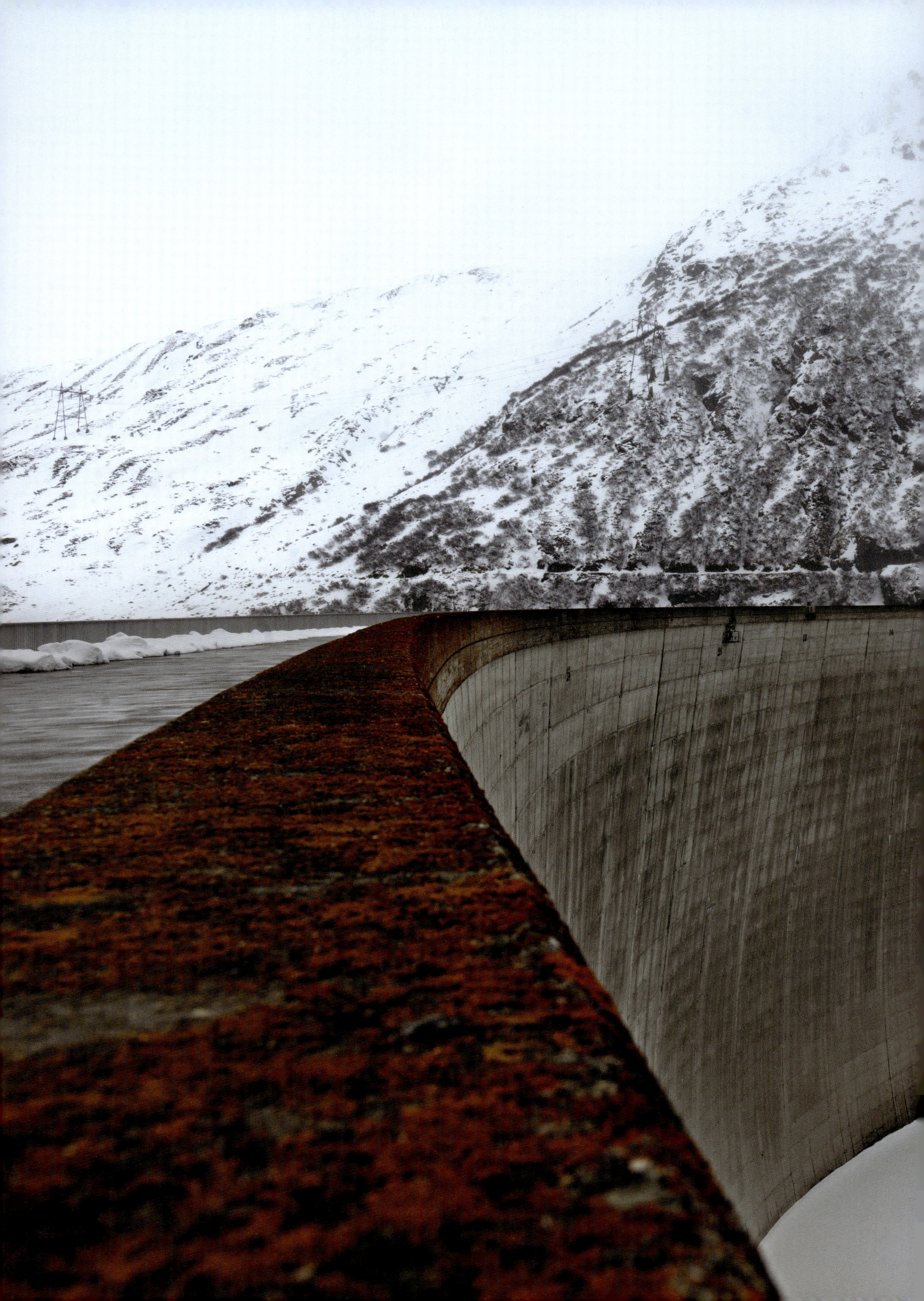

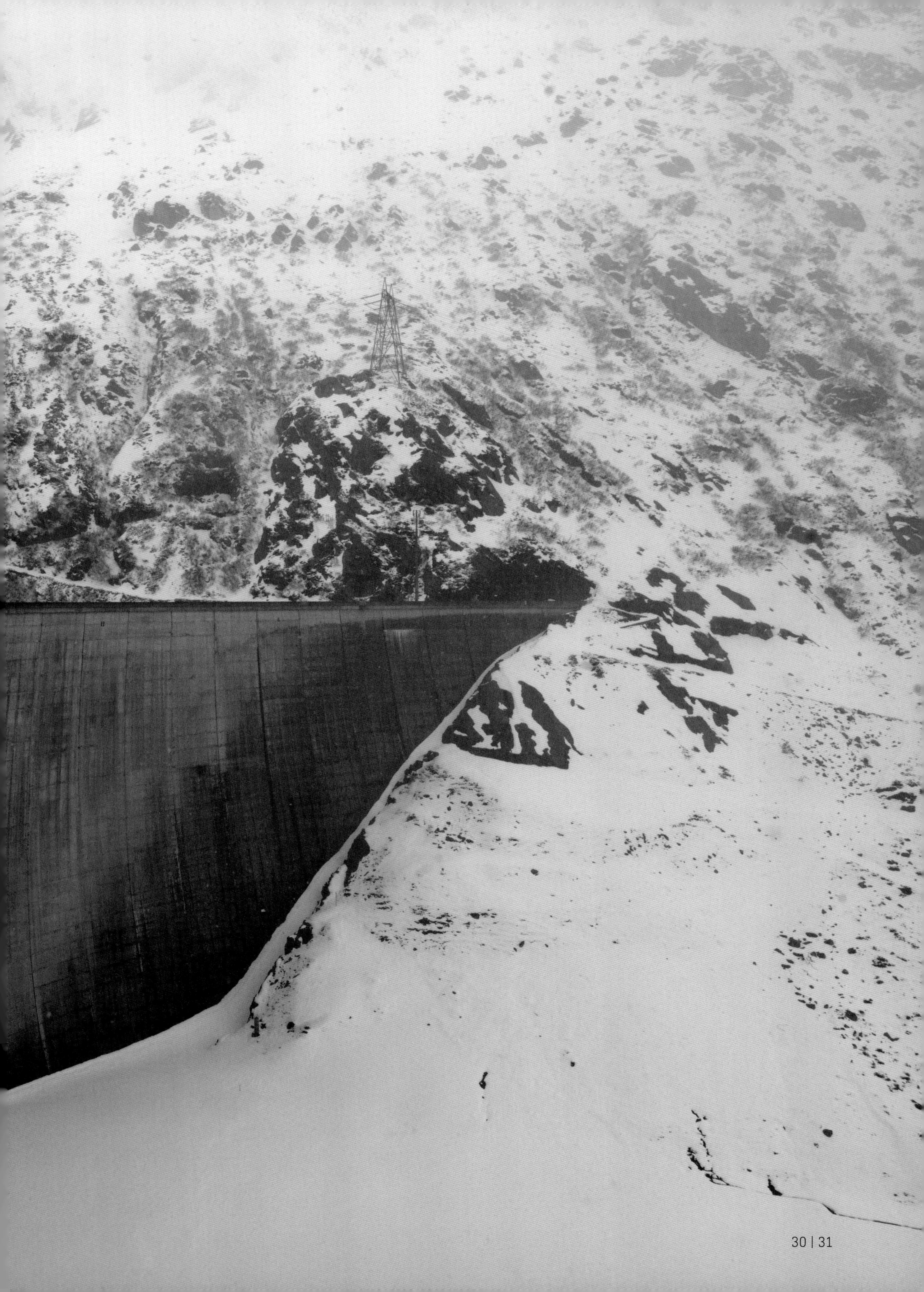

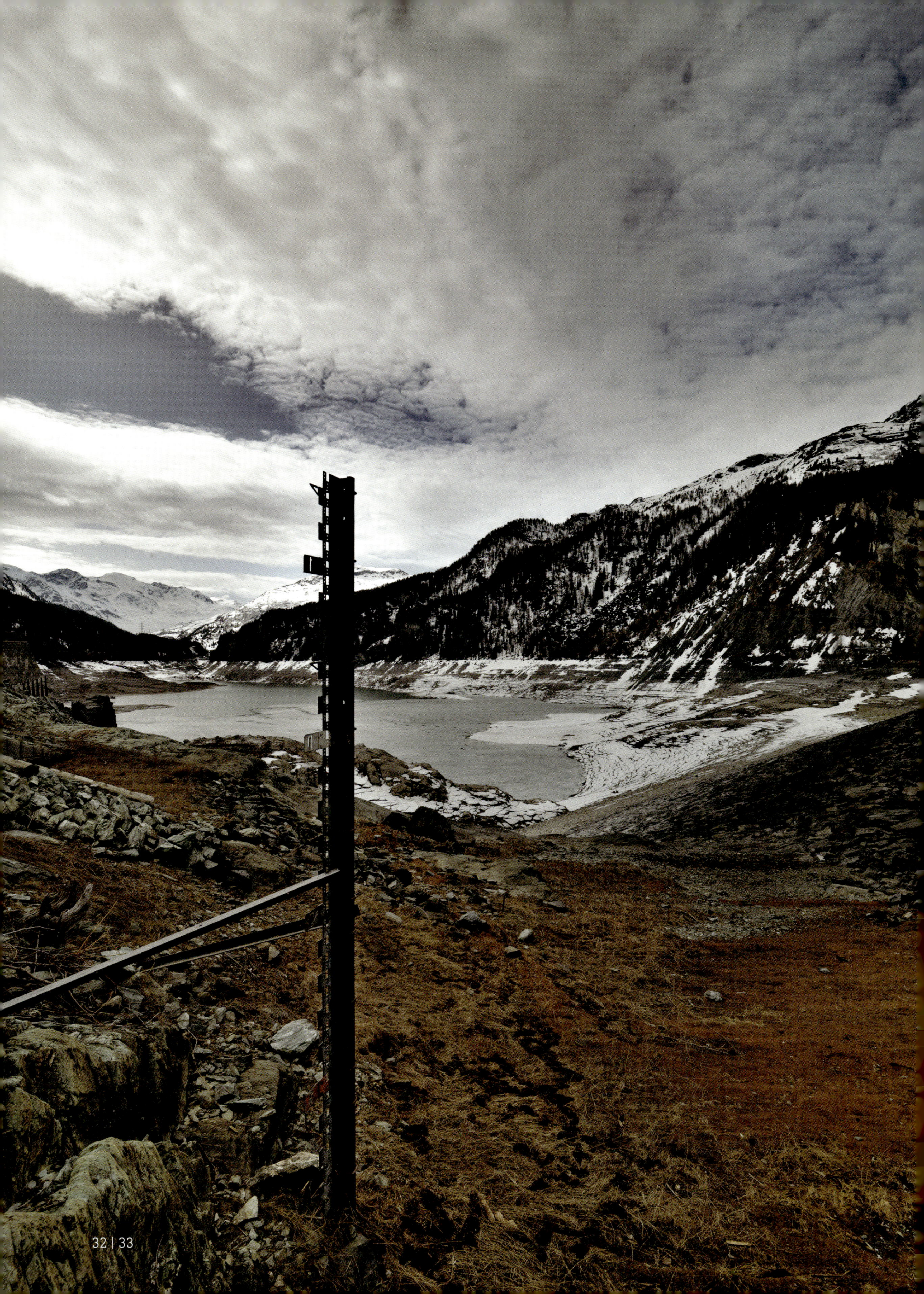

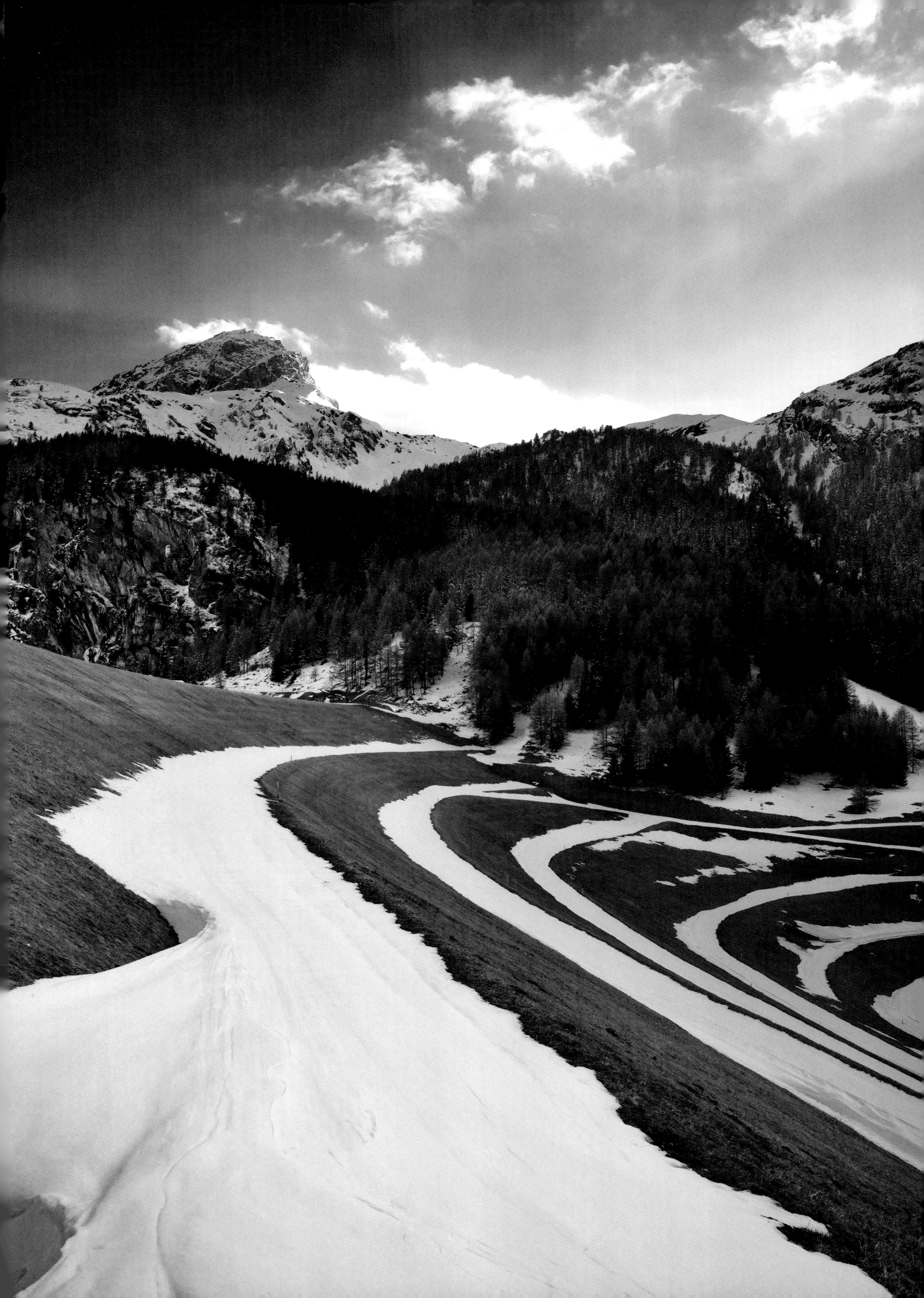

Wilde Wasser, starke Mauern, schöne Erinnerungen

Von Köbi Gantenbein

Die Leute von Marmorera im Oberhalbstein versammelten sich am 8. Mai 1954 ein letztes Mal in der Kirche ihres Dorfes. Sie sangen «Las Mintinedas», ein wunderschönes Bitt- und Klagelied. Die Mintinedas verklangen – Marmorera ging am nächsten Tag unter. Es ruht nun am Boden des Stausees zwischen Savognin und Bivio. Aus ihm holt die Stadt Zürich seit fünfundsechzig Jahren Strom.

Als Kind lebte ich in Samedan im Oberengadin, meine Familie aber kommt aus dem Prättigau. Hier wohnten meine Grosseltern, zu denen ich gerne zu Besuch fuhr. Meistens mit der Eisenbahn durch den Albulatunnel, weiter über Filisur und Davos und dann hinab nach Jenaz. Aufregung war, wenn mein Götti uns besuchte und mich in seinen knallroten VW Käfer setzte. Wir fuhren über den Julier. Nur er und ich. Beim Marmorera See gibt es einen Kiosk. Dort hielten wir an. Wir tranken Bergamotta oder Orangina. Wir assen Mandel- oder Nussgipfel. Wir schauten über den See, wir spazierten über den Damm und mein Götti erzählte mir, dass ein verwunschenes Dorf auf dem Grund des Sees läge – mit Häusern, Ställen und Kirche.

Ich frug ihn, was denn mit den Leuten, den Kühen, den Hunden und Katzen passiert sei. Er zeigte auf die Häuser oben am Hang: Sie sind gezügelt, dort hinauf oder fort ins Unterland. Sie erhielten Kalt- und Warmwasser, eine elektrische Heizung und viel Geld. Viele waren froh und hatten Freude an den neuen Küchen, der Badewanne, der Waschmaschine, dem Kühlschrank, einige hatten nun sogar eine Garage und ein Auto.

Dieses Fortschrittschicksal hat mich als Bub immer wieder beschäftigt. Ich stellte mir vor, wie es sein würde, wenn meine Eltern, mein Bruder Daniel und ich fort müssten aus unserem Haus an der Bahnhofstrasse von Samedan, weil die Seebauer von Marmorera das Engadin unter Wasser setzen würden. Käme Silveli, mein Kindergartenschatz, mit uns? Und Mario, mein Freund – wäre er bei denen, die dorthin fort müssten, was der Götti «Unterland» nannte? Und was wäre mit der dicken Frau Müller oder dem alten Herr Tgetgel. Nähmen wir sie mit? Liessen wir ihn zurück? Von den Katzen meiner Grossmutter wusste ich, dass sie immer selber nach Hause fanden im Herbst, wenn alle von der Alp ins Tal zügelten. Ich stellte mir vor, wie die Katzen von Marmorera tauchend ihr Haus suchen. Und ich sah die unterwasserige Siedlung in geheimnisvoll blauem Licht, so schön wie das Bild meiner Kinderbibel, auf dem Moses das Meer beim Sinai geteilt hatte, damit die Israeliten hindurch schreiten konnten.

Marmorera ist in Zervreila hinter Vals; Marmorera ist in Nalps und Cruvera in der Surselva, ist in Livigno, ist in vierzig Landschaften Graubündens; Marmorera ist in Meinrad Inglins grossartigem Roman «Urwang»; Marmorera ist im Urwald des Itaipu in Paraguay, ist am Jangtsekiang der drei Schluchten, Marmorera ist in Südanatolien, wo für den Ilisu-Stausee am Tigris abertausende Menschen mit Zwang umgesiedelt werden und die uralte Stadt Hasankeyf unterging hinter einem 300 Meter hohen und zwei Kilometer langen Damm, aufgewuchtet auch mit viel Kapital aus der Schweiz.

Nur im Rheinwald, dem Tal zum San Bernardino-Tunnel, ist Marmorera nicht, denn Neu-Splügen wurde nicht gebaut. Schon 1917 hatte sich der Energiekonzern Motor Columbus in Sufers die Konzession für die Wasserkraft am Hinterrhein gesichert. In den Vierzigerjahren gehörte die Schweiz zu den am höchsten elektrifizierten Ländern Europas. Es war absehbar, dass nach Kriegsende der Strombedarf nicht nur im Inland, sondern auch in den Nachbarländern steigen würde. Also wurde kräftig für Stauseen und Kraftwerke Stimmung gemacht. Aber die Rheinwaldner legten sich quer. An einer Landsgemeinde in der Kirche von Splügen rebellierten sie gegen das Projekt. Zum Konsortium, das die Hinterrhein-Werke bauen wollte, gehörten die grossen Energiekonzerne, hinter ihnen die Schweizer Banken. Die stromhungrige Industrie hatte grosse Erwartungen. Dem Entscheid der Regierung des Kantons Graubünden sah man im Winter 1944 deshalb bange entgegen. Aber am 11. März schützte der Kleine Rat des Kantons Graubünden und später der Bundesrat den Entscheid der Einwohner des Rheinwald: Kein Stausee.

Neu-Splügen ist auch am Silsersee im Engadin, in der Val Curciusa, in der Val Madris und auf der Greina. Neu-Splügen war vor dreissig Jahren in Kaiseraugst und ist heute am Wellenberg in Nidwalden, im Zürcher Weinland oder am Oberbauenstock in Uri – nirgendwo werden Atomkraftwerke ihren ewig strahlenden Müll vergraben können. Alle Neu-Splügen werden nicht gebaut, wenn sich genügend Menschen dagegen wehren.

«Roda mulin, ti roda mulin, vas ad incuntin», «Mühlerad, Du Mühlerad, drehe und drehe dich ohne Ende». So dichtete Gian Fontana auf Sursilvan und Tumasch Dolf komponierte daraus ein in Graubünden populäres Kinderlied. Das Mühlerad dreht sich unaufhörlich, es schafft Reichtum. Sein Wasser sprudelt über die Rinne und ruft ins Tal hinunterfliessend: «Lebewohl» – oder eben auf romanisch so schön deutsch: «pietigott».

Was einst in den Alpen das Mühlerad war, war nach dem Zweiten Weltkrieg die Wasserkraftturbine. Georg Peter Luck, Grafiker und Maler, war einer der ersten wichtigen Künstler meines Lebens, denn er hat die Bilder für «Wilde Wasser – Starke Mauern» gemalt, ein Silva Buch von 1962. Den Text schrieb Max Mummentaler. Für die Nachgeborenen – Silva Buch, das war eine Reklame-Idee der Markenartikel-Firmen. Sie druckten auf ihr Banago, Maga & Floris oder die Cailler-Dessert-Schoggi Punkte je nach Warenwert. Die schnitt ich aus, sammelte sie und schickte sie ein. Der Pöstler brachte mir einen Brief mit Bildern. Meine Eltern kauften für wenig Geld das Buch dazu. Dort waren im Lauftext die Stellen ausgespart, wohin ich zusammen mit meinem Vater Hitsch die Bilder klebte. So lernte ich das Drama des Dorfes «Fasano» im Val Verena kennen.

«Fasano» ist Marmorera, ausgeschmückt als Heimat- und Technikroman. Der Gemeindepräsident Pietro Madonna sorgt für Ruhe und Ordnung. Der Pfarrer Don Borella führt seine Schäfchen zum Glauben an den Fortschritt – vor sechzig Jahren waren die Katholiken noch anders, ich las im Buch nirgends, dass der Bischof seinen Priester wegen Fortschrittsfreude geplagt hätte. Ingenieur Markwalder vertritt die Kraftwerkfirma, er erklärt den Fortschritt und verspricht reiche Ernte für alle, er organisiert alles und wird der beste Freund des Gemeindepräsidenten. Dem verspricht er bei Gelingen des Kraftwerks die Stelle als Wirt der Kantine. Und eine Liebesgeschichte gibt es auch. Auf der Baustelle arbeiten mehrere hundert Arbeiter. Für sie gibt es ein eigenes Dorf, ein Spital, einen Laden, ein Kino, eine Bank, einen Coiffeursalon und eine Post. Renzo, der Sohn des Capo, und das Postfräulein bändeln eine Liebelei an und heiraten schliesslich mit einem grossen Fest.

Zuerst sind alle Menschen in der Val Verena gegen das Kraftwerk. Der Pfarrer, der Capo und der Ingenieur drehen alle um. Nur Ephraim Cavalli, der schwarze Ephraim, wehrt sich. Grundsätzlich und weil er sein Schmugglergewerbe gefährdet sieht. Er wird zum Desperado der Alpen. Ich aber ging in der Rolle seines Sohns Fernando auf. Ich wurde Gehilfe des Ingenieurs Markwalder. Ich lernte Land vermessen, ich lernte Geologie, ich liebte den technischen Fortschritt. Nachdem mein Chef Markwalder die Stimmung in der Gemeinde umgedreht hatte, nachdem also Kraftwerk, Stausee und Staumauer froh begrüsst wurden, regierte er eine pharaonische Baustelle. Das Buch erzählt sie anschaulich. Und ich, Fernando, war überall dabei als des Chefs Gehilfe. Ich sass im Kino, eigens gebaut für die hunderte Mineure und Kranführer, ich ass mit ihnen in der Kantine Makkaroni mit Tomatensugo, ich sang mit ihnen Lieder gegen das Heimweh. Ach je – wie hat die grosse Baustelle am Kraftwerk Fernando die Welt geöffnet, wie hat sie also auch mich und das Dorf mit der Welt verbunden. Endlich. Und ich sage gerne und überzeugt, das Kraftwerk hat auch mir, dem Lokiführerbub, Zukunft gegeben. Seine Profite haben geholfen, den Kanton Graubünden an die Moderne anzuschliessen. Sein Geld hat zum Beispiel geholfen, dass die Mittelschulen ausgebaut worden sind und die Unterlandfahrt an die Universität auch für meinesgleichen möglich wurde. Nicht nur, aber auch – die Stromwirtschaft hat den Kanton Graubünden in den Fünfziger- und Sechzigerjahren zu einem normalen Schweizer Kanton gemacht, seine Gesellschaft geöffnet und Luft hereingelassen. Sie hat für diesen Preis nicht nur billigen Strom, sondern auch grossen Profit aus den Bergen ins Unterland gezogen, denn die Stromwirtschaft ist zwar in staatlichem Besitz, aber vielfältig mit privaten Kapitalgebern verwickelt.

Als Bub war ich ab und an zu Besuch in Zillis, wo Angela wohnte, eine Freundin meiner Mutter und Dorflehrerin. Mir fiel natürlich auf, wie viel schöner, grösser und besser eingerichtet ihr Schulhaus war als meines. Wie grossartig war die Schulküche, alles schien es da zu geben für die Kochschule; einen Elektroherd zum Beispiel mit einem Schublädchen unter den Heizplatten. Lief etwas über, so rann es da hinein. Und in der Turnhalle schliesslich standen die neuesten Turngeräte von Adler + Eisenhut. In Malans, wohin meine Familie aus dem Engadin mittlerweile gezügelt war, hatten wir weder eine Werk- noch eine Kochschule, kaum eine Turnhalle – denn die Ziraner hatten etwas, was wir nicht hatten – den Wasserzins, die die Kraftwerke für ihre Kilowattstunden abgeben mussten.

In ein paar Jahren werde ich mich als Nachfolger von Mummenthaler und Luck daran machen, den zweiten Band von «Wilde Wasser – Starke Mauern» zu zeichnen und zu schreiben. Denn das Publikum will wissen, wie es mit Fernando, Ingenieur Zellweger und allen weitergegangen ist. Ich werde berichten, wie dramatisch die Ausbeutung der Bäche und Flüsse die Landschaft, wie tiefgreifend das Ertrinken des Val Verena die Gesellschaft Graubündens verwandelt hat. Renzo, der Sohn des Capo, der das Postfräulein geheiratet hat, wird in meinem Fortsetzungsband sozialdemokratischer Regierungsrat werden. Er wird in der Arena des Schweizer Fernsehens sitzen und die Besserwisser, die seinen Kanton als alpines Armenhaus verlachen, mit ein paar Tatsachen deckeln. «Die Fakten, mein Herr», wird er in die Kamera rufen, «haben Sie vergessen. Graubünden produziert 8000 Millionen Kilowattstunden Elektrisch im Jahr. Das sind mehr als ein Fünftel der Energie der Schweiz. 40 Kraftwerkgesellschaften, die 55 Landschaften mit Staumauern unter Wasser gesetzt haben, sorgen dafür, dass euch im metropolitanen Raum die Lichter für eure klugen Gedanken nicht ausgehen. Wir aber lassen es gar grosszügig gut sein mit etwas Wasserzins. Die Steuersitze der Gesellschaften aber», wird Renzo mit rotem Kopf wettern, «haben über all die Jahre im Unterland fetteste Gewinne abgetragen, wir werden uns damit nicht mehr zufrieden geben.» Und sie streiten ergebnislos noch lange.

Ich werde aber auch staatstragend dichten, so wie Mummenthaler und Luck im ersten Band. Ich werde nachzeichnen, wie im Kanton Graubünden gut 500 Arbeitsplätze entstanden sind für Schlosser, Elektriker, Maurer, Schreiner oder Informatiker, die den Unterhalt der Werke besorgen. Dazu Verwalterinnen, Stromhändler, Ingenieure, Buchhalterinnen und Unter- und Oberchefs. Und hunderte Lieferanten vom Metzger für die Bratwürste der Verwaltungsratsitzungen bis zu den Lehrerinnen für die Kraftwerkkinder. Wir werden Bilder sehen, wie traute Familien in Safien im Stubenlicht aus Gratisstrom Znacht essen und ihr Zimmer fast gratis elektrisch heizen. Ein Kapitel wird nachzeichnen, wie Wasserzinsgemeinden den Geldsegen verschwenden, um über Jahrzehnte die Steuern absurd tief zu senken. Was natürlich keine Entwicklung bringt. Und im anderen Kapitel sehen wir das Gegenteil: Maurizio, der zweite Sohn des schwarzen Ephraim, zieht nach Vals. Er wird dort Hotelier und hat mit einer Gruppe Aufgeweckter mit dem Wasserzins-Geld aus Zervreila das verkommene Hotel Therme wachgeküsst dank eines Badbaus von Peter Zumthor. Und auch das Drama werde ich in bunten Farben ausmalen, wie die Valser den Fortschritt aus dem Wasserzins eine Generation später verdummt und vernichtet haben, die Therme einem mysteriösen Mann des Geldes überlassend.

Der Strom war die wichtigste technische, kulturelle und gesellschaftliche Veränderung der Schweiz im 20. Jahrhundert. In meinem jungen Lebensfaden habe ich seine Wirkung unmittelbar erlebt. Im Sommer war ich als Knecht bei einem Onkel meiner Mutter in Furna auf Saison. Meine Erinnerung ist idyllisch und überformt von Schönheit. Melken von Hand, Holz sägen auch, Kochen am Feuer – Öhi Michel war ein wendiger Feuerkoch und Flammensenn. Bäsi Maria, eine kluge und schöne Frau, lebte mit der Sonne. Viel erfindungsreiche Achtsamkeit am Tag und geheimnisvolles Leuchten aus der Petrollampe am Abend und eine dunkle Nacht. Schön vor allem, wenn es nicht regnete, und wohl auch im Wissen, dass ich im Herbst wieder ins Tal konnte. Die Furner hatten Angst vor dem Strom, denn 1923 hatte eine Pleite der Bündner Kraftwerke die kleine Gemeinde ruiniert. Ausserdem war man auch skeptisch wegen der Feuerfäden in der Glühlampe und wollte das Geld der Gemeinde lieber für einen Alpstall eingesetzt wissen. Doch dann war es so weit – ich war leider nicht dort, liess es mir aber immer wieder erzählen: Das elektrische Licht und der Strom aus der Steckdose kamen auch in Furna an, das letzte stromlose Dorf in Graubünden begann 1968 zu leuchten. Mein Öhi Michel war glücklich samt Bäsi Maria – endlich Strom.

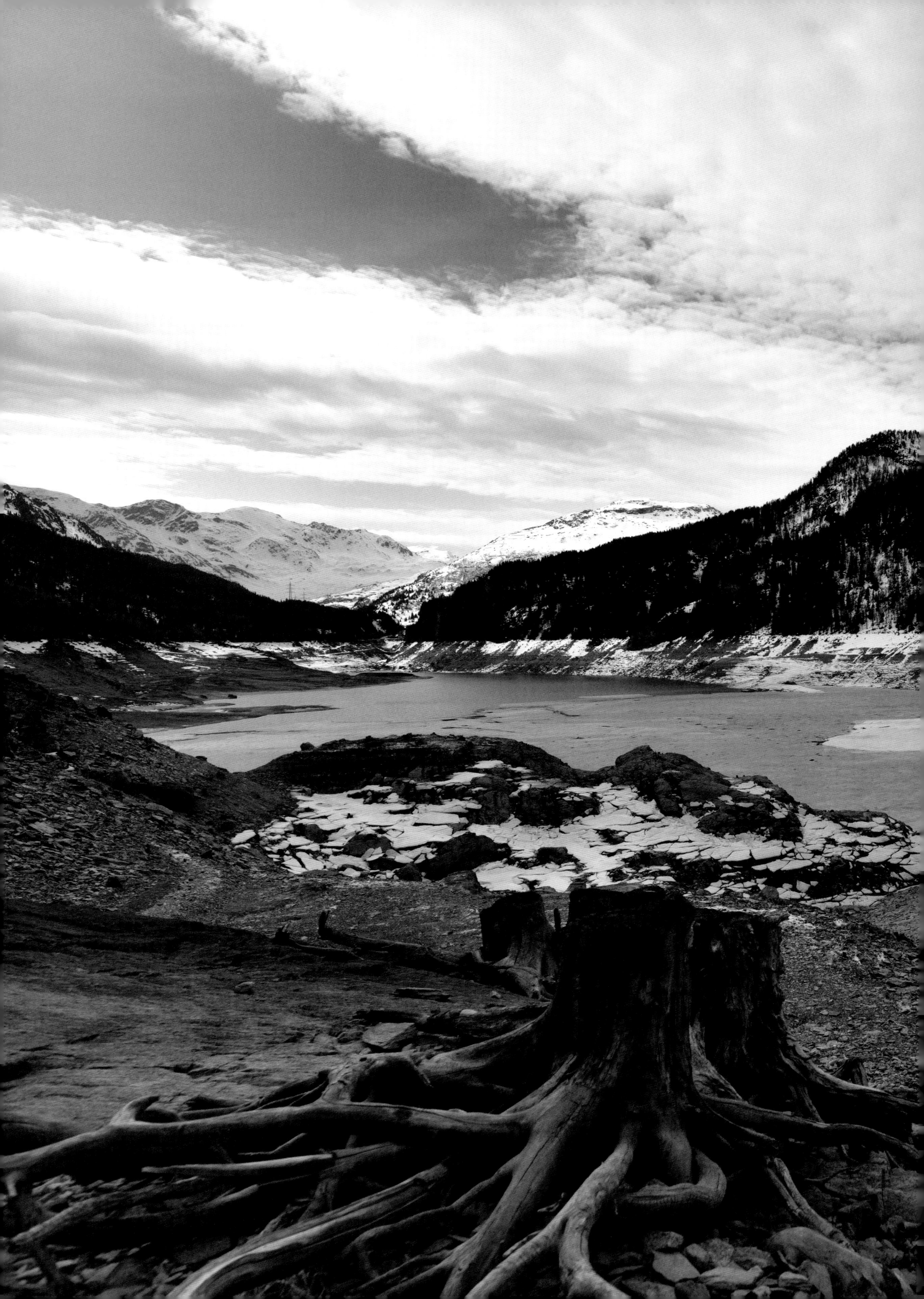

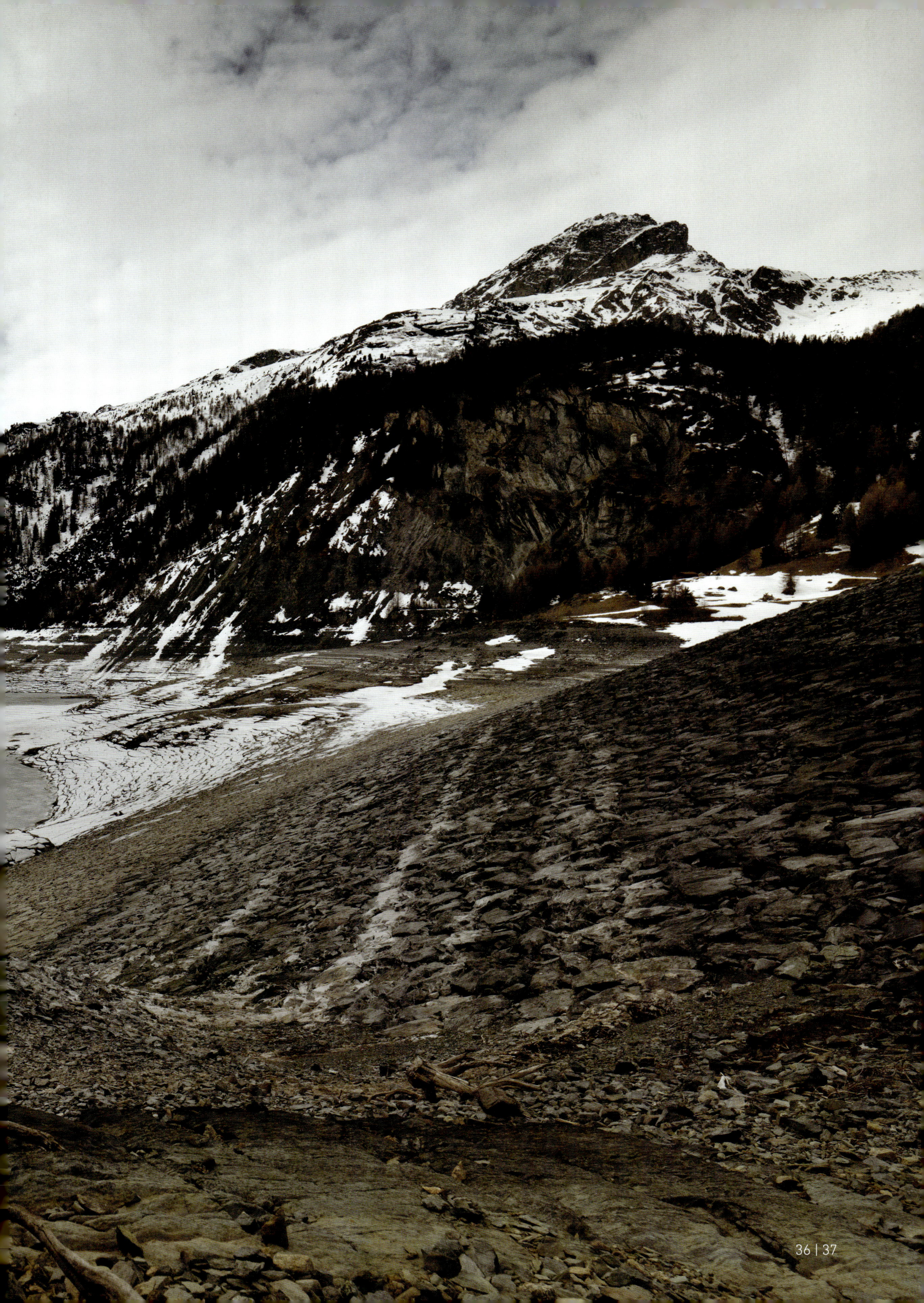

Wild Waters, Strong Walls, Beautiful Memories

By Köbi Gantenbein

The residents of Marmorera in the Surses region in Switzerland gathered for the last time in their village church on May 8, 1954. They sang "Las Mintinedas", wonderful songs of solicitation and lament. The Mintinedas were over – and Marmorera disappeared the next day. It now rests on the bottom of the reservoir lake between Savognin and Bivio, which has been supplying the city of Zurich with electricity for the past 65 years.

As a child I lived in Samedan in the Upper Engadine region, but my family hails from the Prättigau. This is where my grandparents lived whom I loved to visit. I usually went by train through the Albula Tunnel, and on past Filisur and Davos and down to Jenaz. It was quite a thrill when my godfather visited and let me ride in his bright red VW beetle. We drove across the Julier Pass. Just him and me. Near Lake Marmorera there is a kiosk. This was where we used to stop. We drank Bergamotta or Orangina. We ate almond or nut crescents. We looked across the lake and walked across the dike and my godfather told me that an enchanted village was at the bottom of the lake – complete with houses, stables and a church.

I asked him what happened to the people, the cows, dogs and cats. He pointed to the houses on the slope – they moved up there or to the lowlands. They received cold and warm tap water, electric heating and a great amount of money. Many were glad and enjoyed the new kitchens, the bathtub, the washing machine and the refrigerator, some even had a garage and a car.

As a young boy, the destiny of this village resulting from progress kept me thinking. I imagined what it would be like if my parents, my brother Daniel and I were forced to move away from our house on the Bahnhofstrasse of Samedan because the lake builders of Marmorera decided to flood the Engadine. Would Silveli, my Kindergarten sweetheart, move with us? And my friend Mario, would he be among those who would have to move to what my godfather called "lowlands"? And what about the portly Mrs. Müller and old Mr. Tgetgel. Would we take her along? Would we leave him behind? I knew that my grandmother's cats always found their own way back in the fall when everyone moved from the alp to the valley. I imagined the cats of Marmorera diving to look for their homes. And I pictured the settlement under water bathed in a mysterious blue light, as beautiful as the picture in my children's bible in which Moses divided the sea near Sinai for the Israelites to pass through.

Marmorera's destiny can be found in Zervreila behind Vals; it can be found in Nalps and Cruvera in the Surselva district, it can be found in Livigno and in forty landscape of Grisons; Marmorera's destiny is also found in Meinrad Inglin's great novel "Urwang"; it is found in the rainforest of Itaipu in Paraguay and in the three gorges of Jangtsekiang, it is found in South Anatolia, where thousands of people were forced to move for the Ilisu Dam on the Tigris and the ancient city of Hasankeyf sank behind a 300-meter-high and two-kilometer-long dam, a huge project built also with a large amount of Swiss capital.

Only in the Rheinwald, the valley of the San Bernardino Tunnel, Marmorera's destiny is not shared, because Neu-Splügen was never built. As early as 1917 the electric power company Motor Columbus secured the concession for the hydropower of the Hinterrhein region. In the 1940s, Switzerland was among Europe's most electrified countries. It was expected that after the war, the demand for electricity would increase not only inside Switzerland, but also in neighboring countries. Therefore, dams and power plants were intensely promoted. But the residents of Rheinwald refused. They rebelled against the project. In a country parish in the church of Splügen they rallied against the project. The consortium that wanted to build the Hinterrhein power plant included the large electricity companies supported by the Swiss banks. The power-hungry industry had high expectations. Therefore, the local residents anxiously awaited the decision of the government of the Grisons canton in the winter of 1944. But on March 11, the local government of Grisons and later the national government supported the decision of the Rheinwald residents: no dam was built.

Neu-Splügen was repeated at Lake Sils in the Upper Engadine, at Val Curciusa, at Val Madris and at the Greina. Thirty years ago, Neu-Splügen was repeated in Kaiseraugst and today in Wellenberg in Nidwalden, in the Zurich wine country, or near Oberbauenstock in the canton Uri – nuclear power plants have nowhere to bury their eternally radiating nuclear waste. All Neu-Splügen projects are not implemented if enough people oppose them.

"Roda mulin, ti roda mulin, vas ad incuntin", "Mühlerad, Du Mühlerad, drehe und drehe dich ohne Ende (Mill wheel, you mill wheel, turn and turn endlessly)". Gian Fontana of Sursilvan wrote the poem and Tumasch Dolf composed a children's song based on it that is popular in Grisons. The mill wheel turns endlessly, it creates wealth. Its water gushes over the trough and as it flows into the valley it cries "farewell" – or in the local slang: "pietigott".

After World War II, the mill wheel was replaced by the hydropower turbine in the Alps. Georg Peter Luck, graphic designer and painter, was one of the first important artists in my life because he drew the pictures for "Wilde Wasser – Starke Mauern" (Wild waters, strong walls), a Silva book of 1962. The text was written by Max Mummentaler. For those born later or outside Switzerland – Silva books were an advertising gimmick of brand companies. They printed points on their Banago, Maga & Floris or Cailler chocolate, depending on the value of the product. I would cut them out and send them in. The mailman brought me a letter with pictures and my parents bought the associated book at a low price. In the book there were empty areas in the text where I glued the pictures together with my father Hitsch. This is how I became familiar with the drama of the village of "Fasano" in Val Verena.

"Fasano" is actually Marmorera in the guise of a "Heimat" and technology novel. The president of the commune, Pietro Madonna is in charge of law and order. The priest Don Borella guides the faithful towards the belief in progress – sixty years ago, Catholics were different, I did not read anywhere in the book that the bishop scolded his priests for loving progress. Engineer Markwalder represents the electrical power company, he explains the progress and promises a bountiful harvest for all, he organizes everything and becomes the best friend of the president of the commune. He promises him the post of canteen manager if the power plant materializes. And there is also a love story. There are several hundred workers employed at the construction site. They have their own village, hospital, store, a bank, a hairdresser and a post office. Renzo, the son of the village chief, and the young lady working at the post office fall in love and finally get married with a big celebration.

At first all residents of Val Verena are opposed to the power plant. The priest, the president, and the engineer turn everyone around. Only Ephraim Cavalli, the black Ephraim, resists. As a matter of principle and he feels that his contraband business is threatened. He becomes the desperado of the Alps. For my part, I identified with the role of his son Fernando. I became the assistant of the engineer Markwalder. I learned how to chart terrains, I learned geology, I loved technical progress. Once my boss Markwalder had turned the attitude of the community around, i.e. after the power plant, reservoir lake and retaining wall were gladly accepted, he ruled a Pharaonic construction site. The book described it very vividly. And I, in the role of Fernando was involved everywhere as the assistant of the boss. I sat in the cinema especially constructed for the hundreds of miners and crane operators, I ate with them macaroni with tomato sauce, I sang with them songs about homesickness. Oh my, how the great construction site of the power plant opened the world to Fernando, and how it therefore also connected me and the village to the world. Finally. And I gladly and wholeheartedly say it, the power plant also gave me, the son of the locomotive driver, a future. Its profits helped catapult the Grisons canton to modernity. Its money helped, for example, to develop secondary schools and travel to universities in the lowlands also became possible for the likes of me. Not exclusively, but partially also due to the electric power sector development, the Grisons canton became a normal Swiss canton in the 1950s and 1960s, opened up its society, and let fresh air in. In return it not only provided inexpensive electricity but also great profit from the mountain to the lowlands, because even though the electrical power sector is owned by the state, it involves private investors in various ways.

As a boy I would every once in a while visit Zillis, the home of Angela, a friend of my mother's and village teacher. Of course I noticed how much more attractive, bigger and better equipped their school building was than mine. How terrific the school kitchen was, it seemed to contain everything for teaching cooking. For example an electric stove with a small drawer under the heating plates. If something spilled over it flowed into it. And in the gymnasium finally, there was the latest sports equipment from Alder + Eisenhut. In Malans, where my family had in the meantime moved from the Engadine, we had neither arts nor cooking facilities, barely a gymnasium, because the residents of Zillis had something we lacked – the water fee that the power plants had to pay for their kilowatt hours.

In a few years I will succeed Mummentaler and Luck and draw and write the second volume of "Wilde Wasser – Starke Mauern". Because the audience wants to know what happened next to Fernando, engineer Zelleweger and all the others. I will describe how dramatically the exploitation of brooks and rivers has changed the landscape and how fundamentally the drowning of Val Verena affected the Grisons society. In my continuation of the story, Renzo the son of the president who married the post office lady, will be a social democrat senior civil servant. He will sit in the arena of Swiss television and bombard the know-it-alls who make fun of his canton as an alpine poorhouse with a few facts. "Ladies and gentlemen", he will cry into the cameras, "you are missing the fact that Grisons produces 8000 million kilowatt hours of electricity per year. That is more than one-fifth the total energy output of Switzerland. 40 power companies that have flooded 55 landscapes with retaining walls, make sure that the lights do not go out on your brilliant ideas in your urban environment. We generously make do with a little water fee. However, the controlling headquarters of the companies", Renzo will go on heatedly "have made the greatest profit all these years in the lowlands, we will no longer be satisfied with this". And they will continue to argue pointlessly for a long time.

But my fiction will also extend to the economic situation on location, similar to Mummentaler and Luck in the first volume. I will describe how a good 500 jobs were created in the Grisons canton for welders, electricians, builders, carpenters or computer specialists, who are in charge of maintaining the plants. Add to that administrators, electricity brokers, engineers, bookkeepers and middle and upper management. And hundreds of suppliers from the butcher delivering hundreds of sausages for management board meetings up to teachers for the children of power plant employees. We will see pictures how happy families in Safien have supper in rooms illuminated by free electricity and heated by electricity costing next to nothing. One chapter will describe how water fee communities waste the financial bonanza by absurdly lowering taxes. Which naturally prevents development. While in another chapter we will see the opposite: Maurizio, the second son of the black Ephraim, moves to Vals. He becomes a hotel owner and with a group of entrepreneurial friends uses the water fee money of Zervreila to reawaken the Hotel Therme with a spa addition by Peter Zumthor. I will also vividly present the drama how one generation later the residents of Vals will have wasted and destroyed the progress based on the water fees and hand over the spa to a mysterious man of means.

Electricity was the most important technical, cultural and social change factor of Switzerland in the 20th century. In my younger years I have directly experienced its effects. I worked as a farm help during the summer season for an uncle of my mother's in Furna. My memories are idyllic and dominated by beauty, hand milking, wood cutting, and cooking on open fire – uncle Michel was a skilled open fire cook and alpine dairyman. Aunt Maria, a smart and beautiful woman, lived with the sun. Plenty of creative attentiveness during the say and the mysterious light of the kerosene lamp by night. It was exceptionally beautiful when it was not raining and probably also due to the knowledge that in the fall I would return to the valley. The residents of Furna were afraid of electric power because in 1923 the bankruptcy of the Bünden power plant had ruined the small commune. They were also skeptical of the glowing filaments in light bulbs and preferred to use the money of the commune for a common stable. But then the time came – I was not there, unfortunately, but enjoyed listening to the story over and over: Electric light and power sockets arrived in Furna and the last village in Grisons without electric power was illuminated in 1968. My uncle Michel was happy, and so was aunt Maria – they finally had electricity.

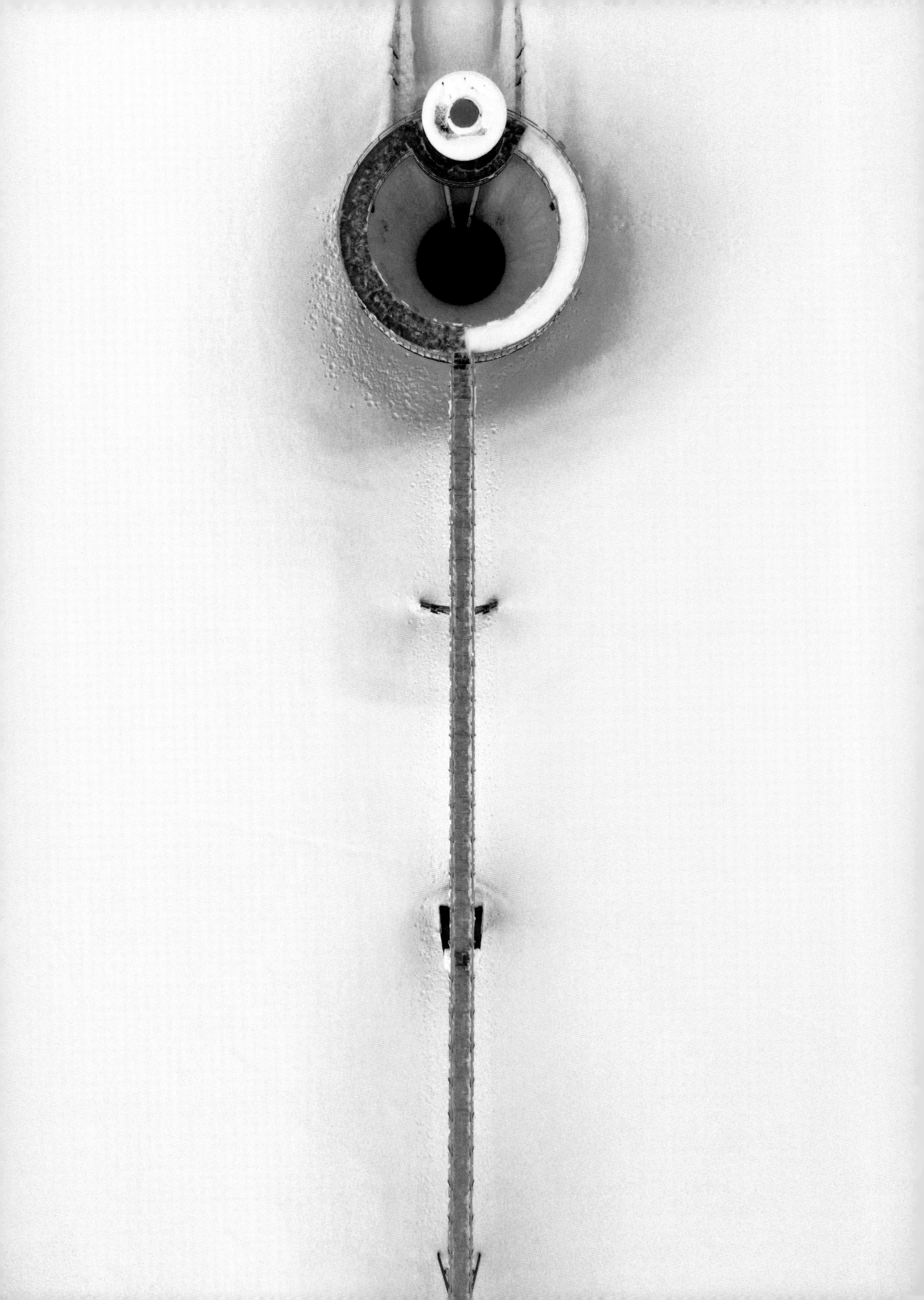

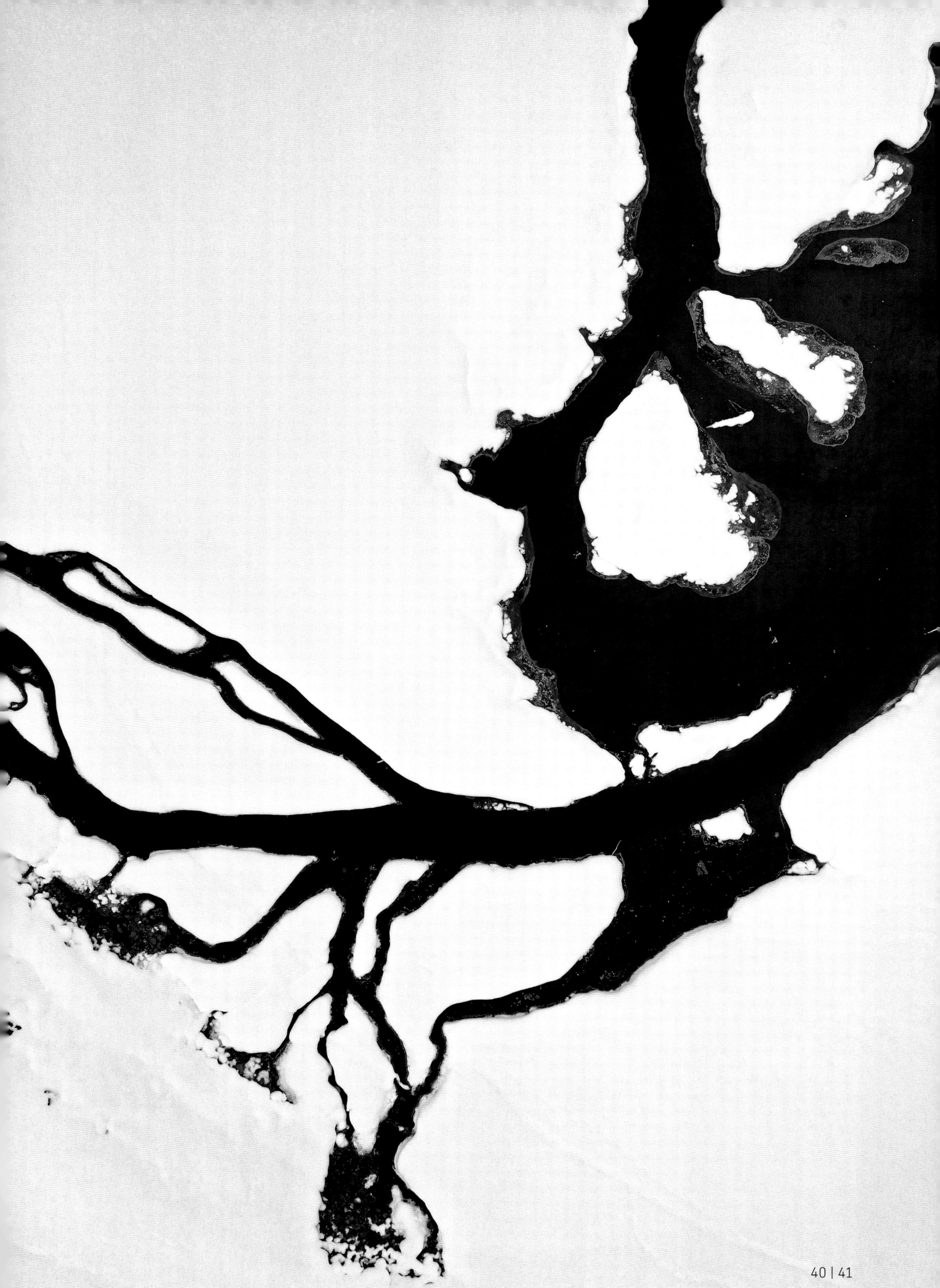

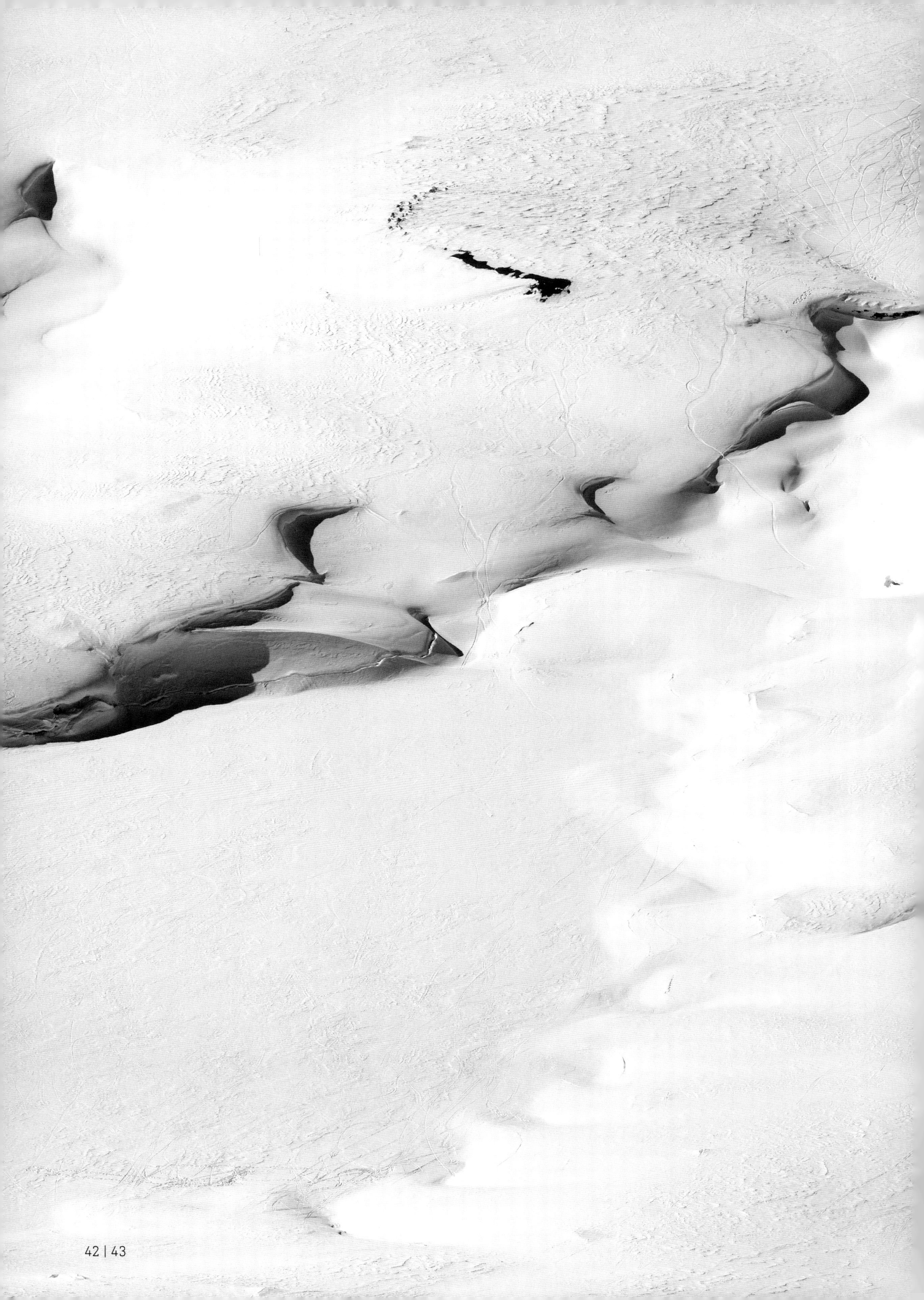

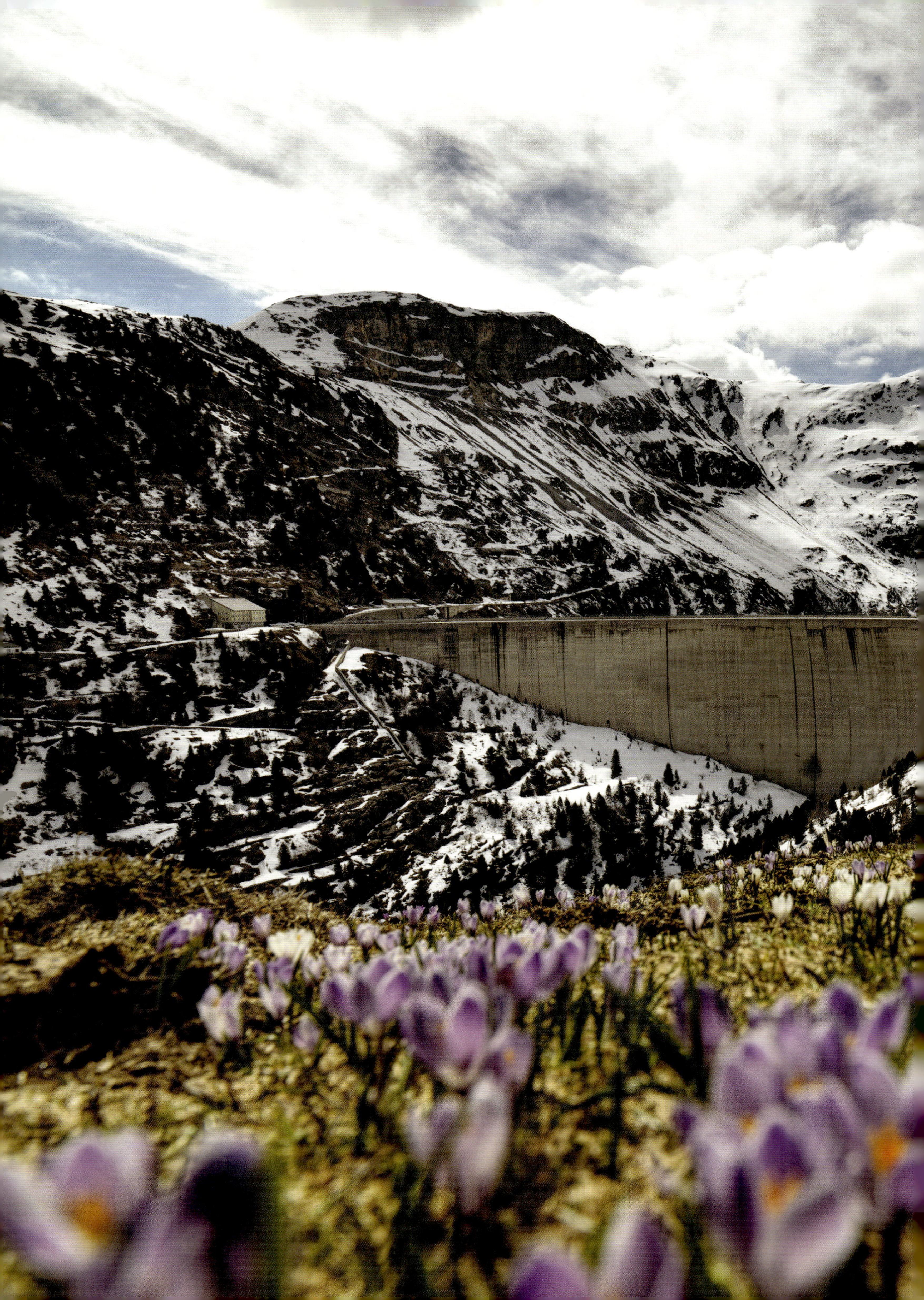

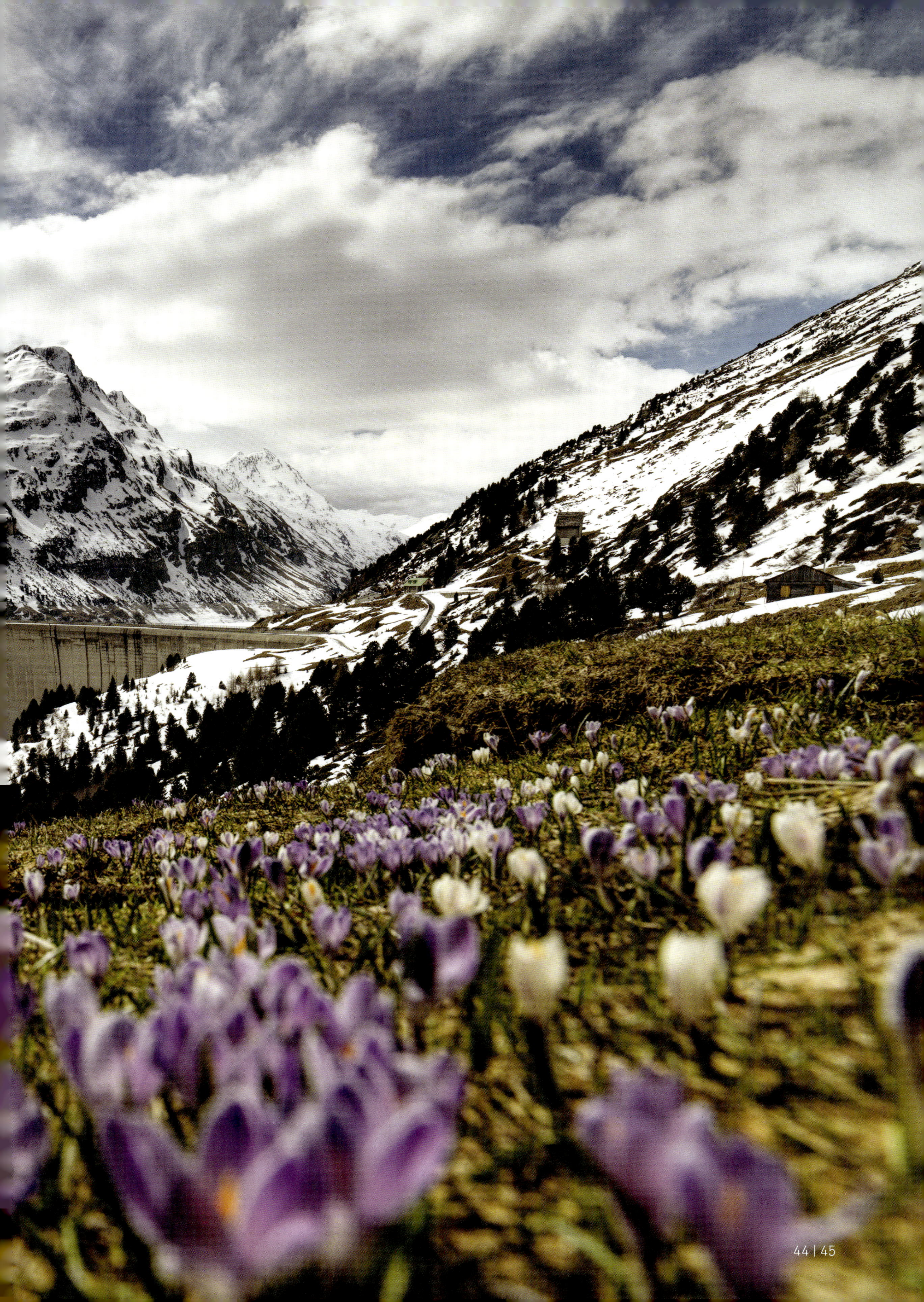

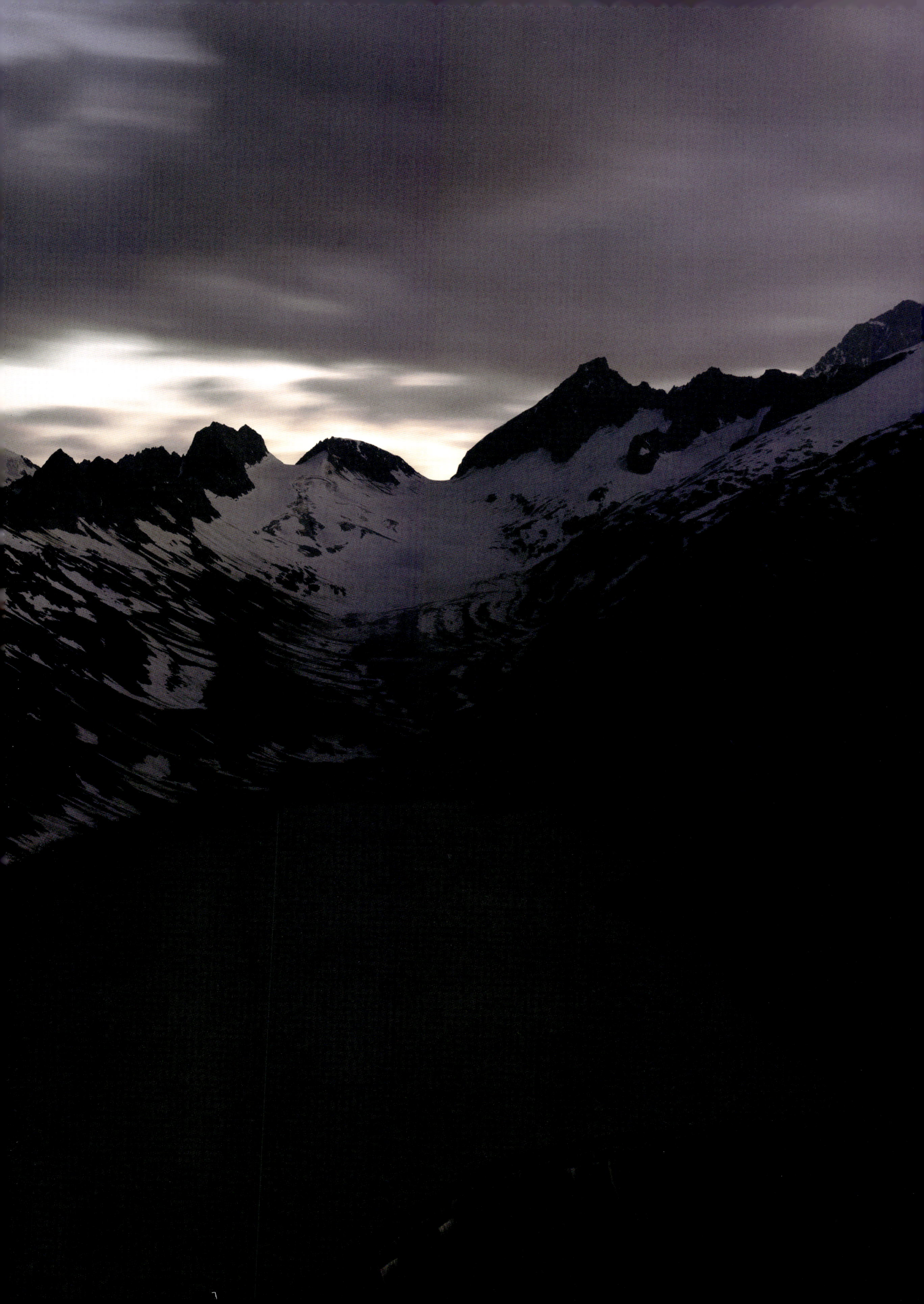

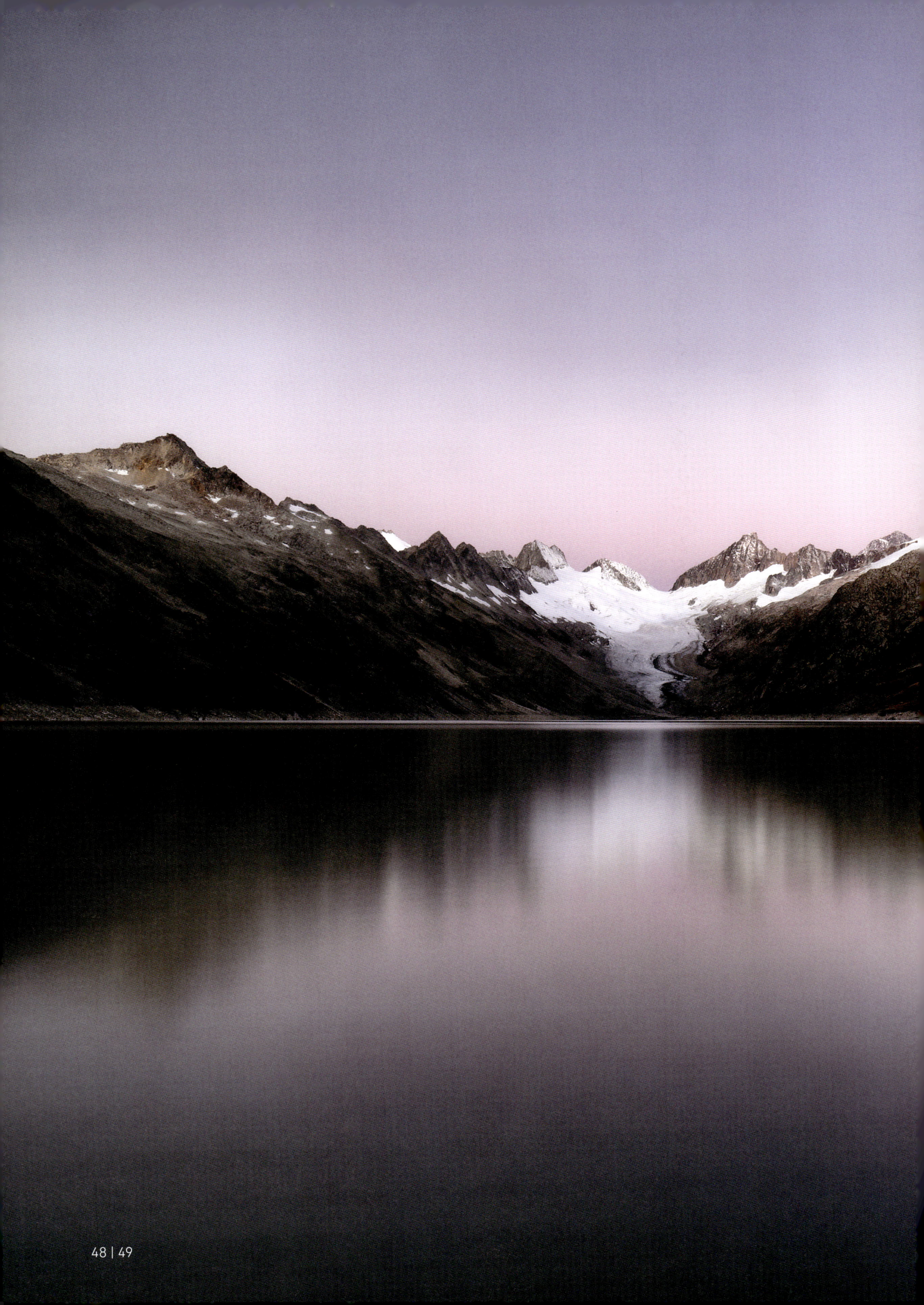

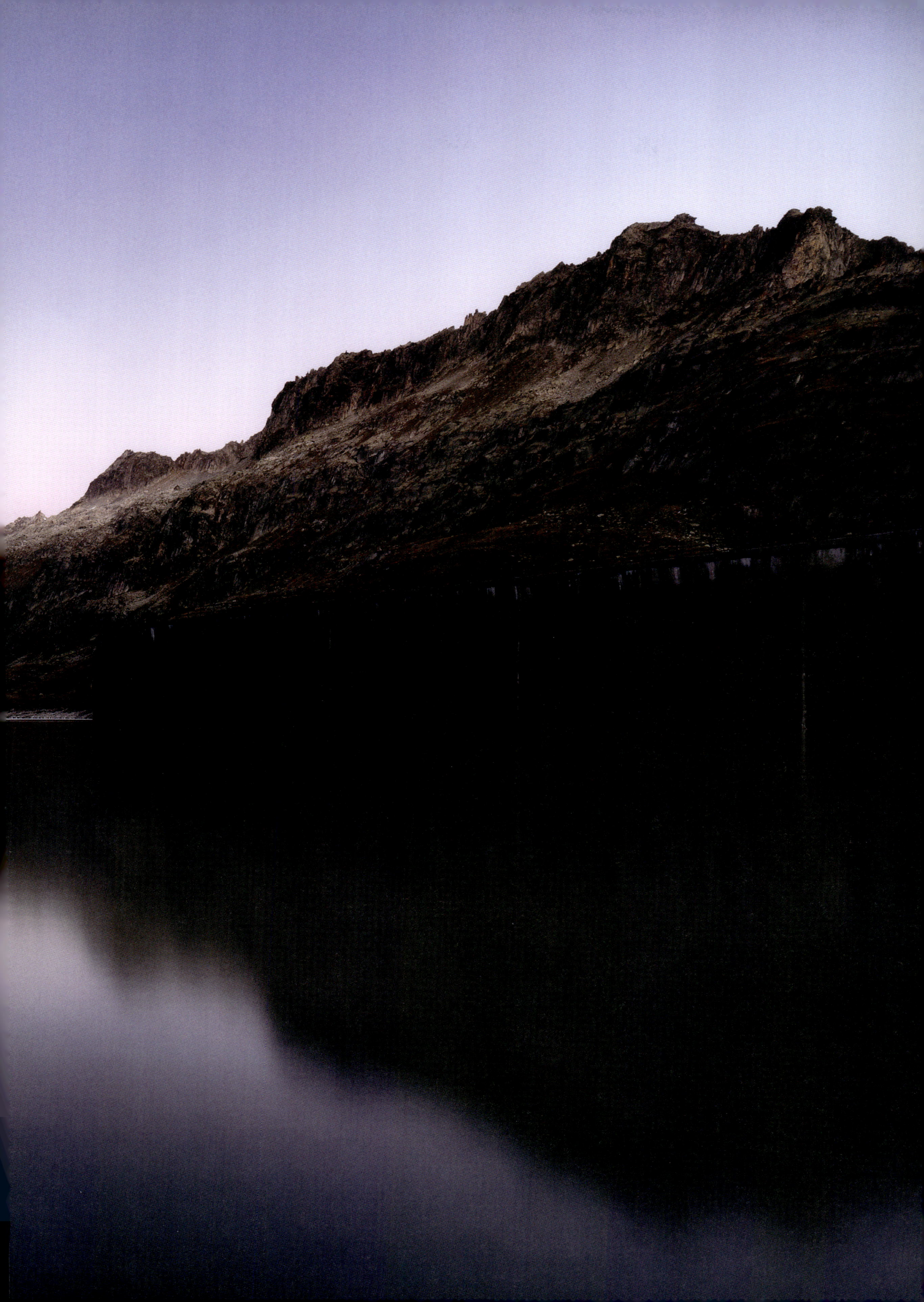

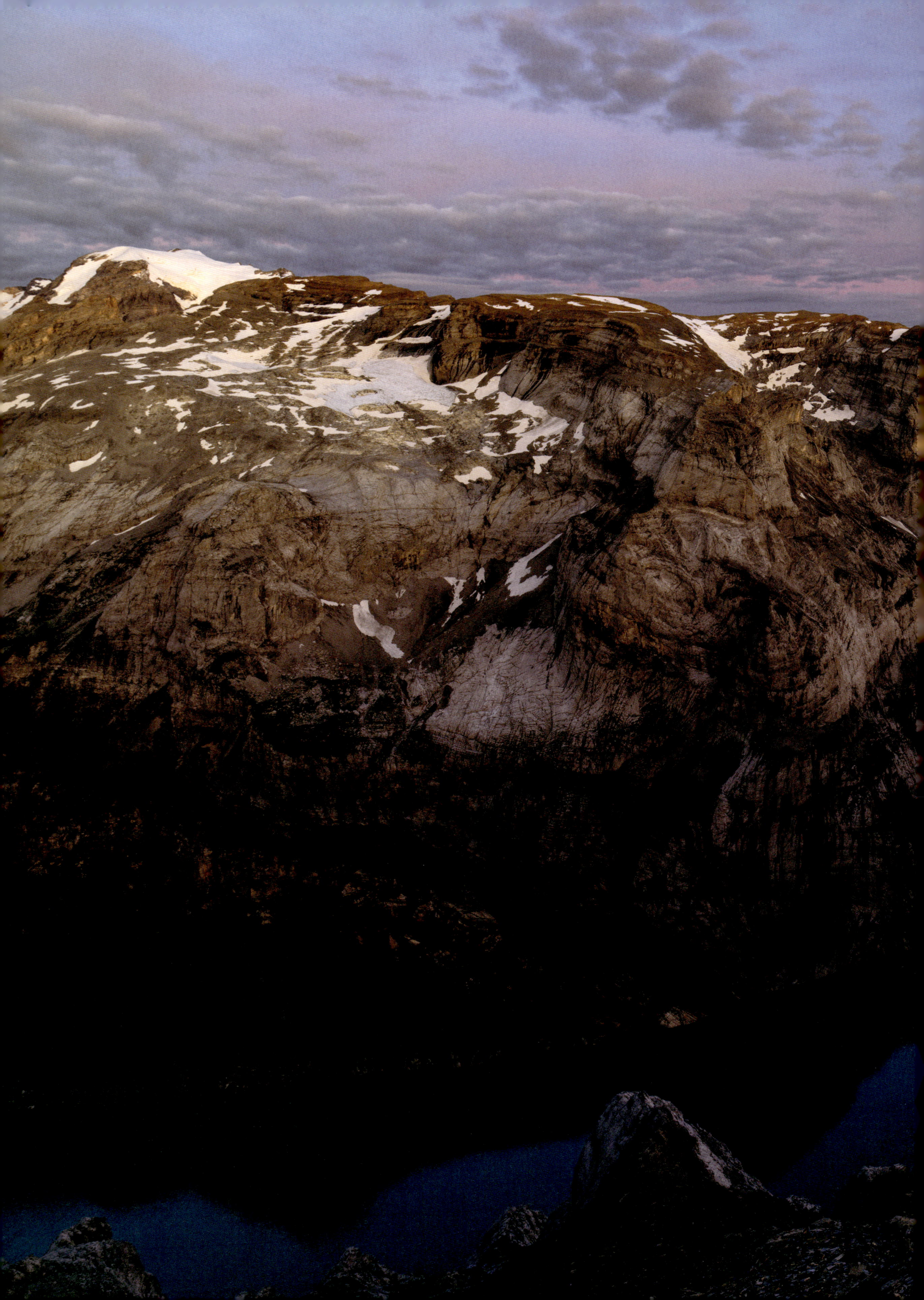

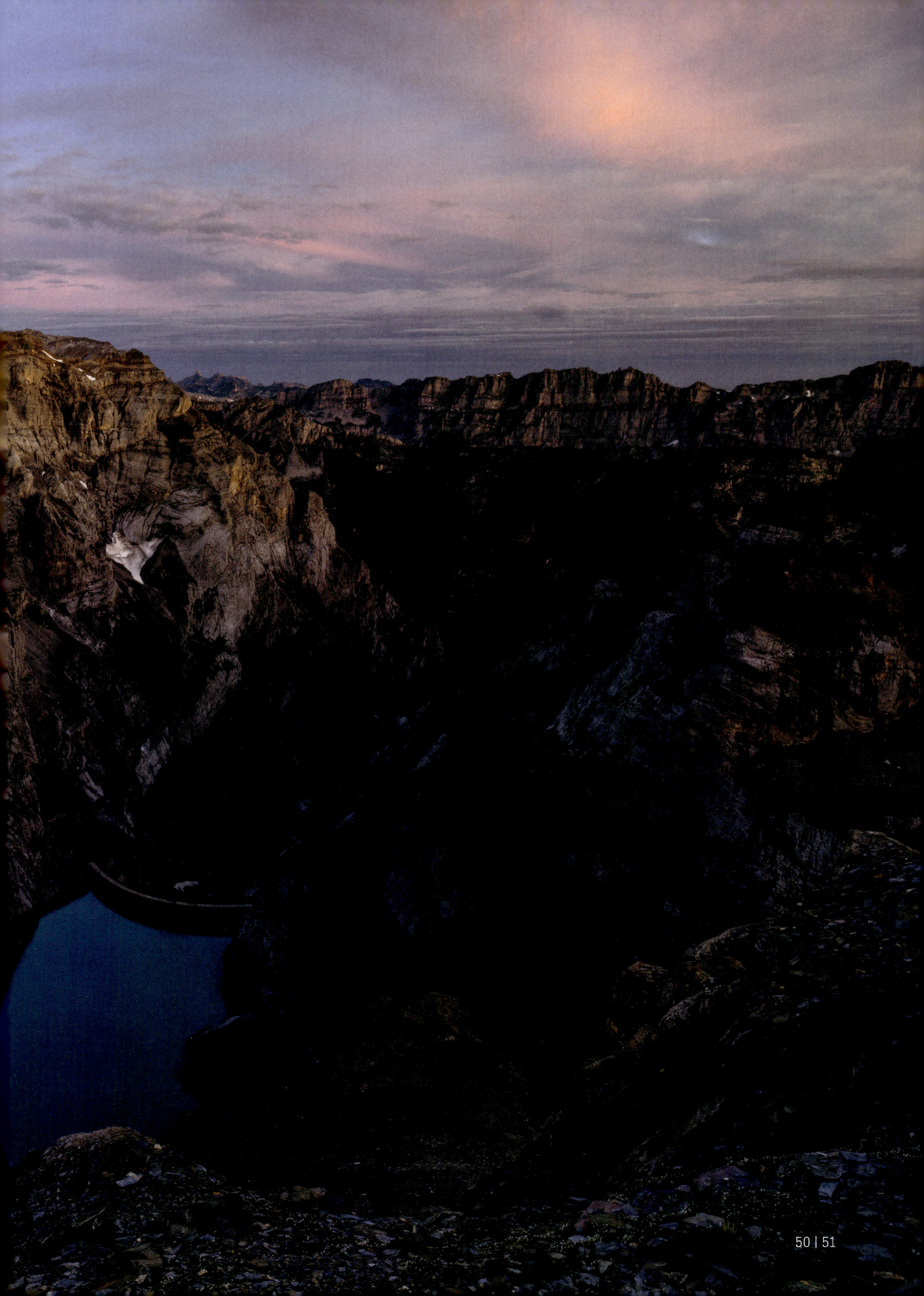

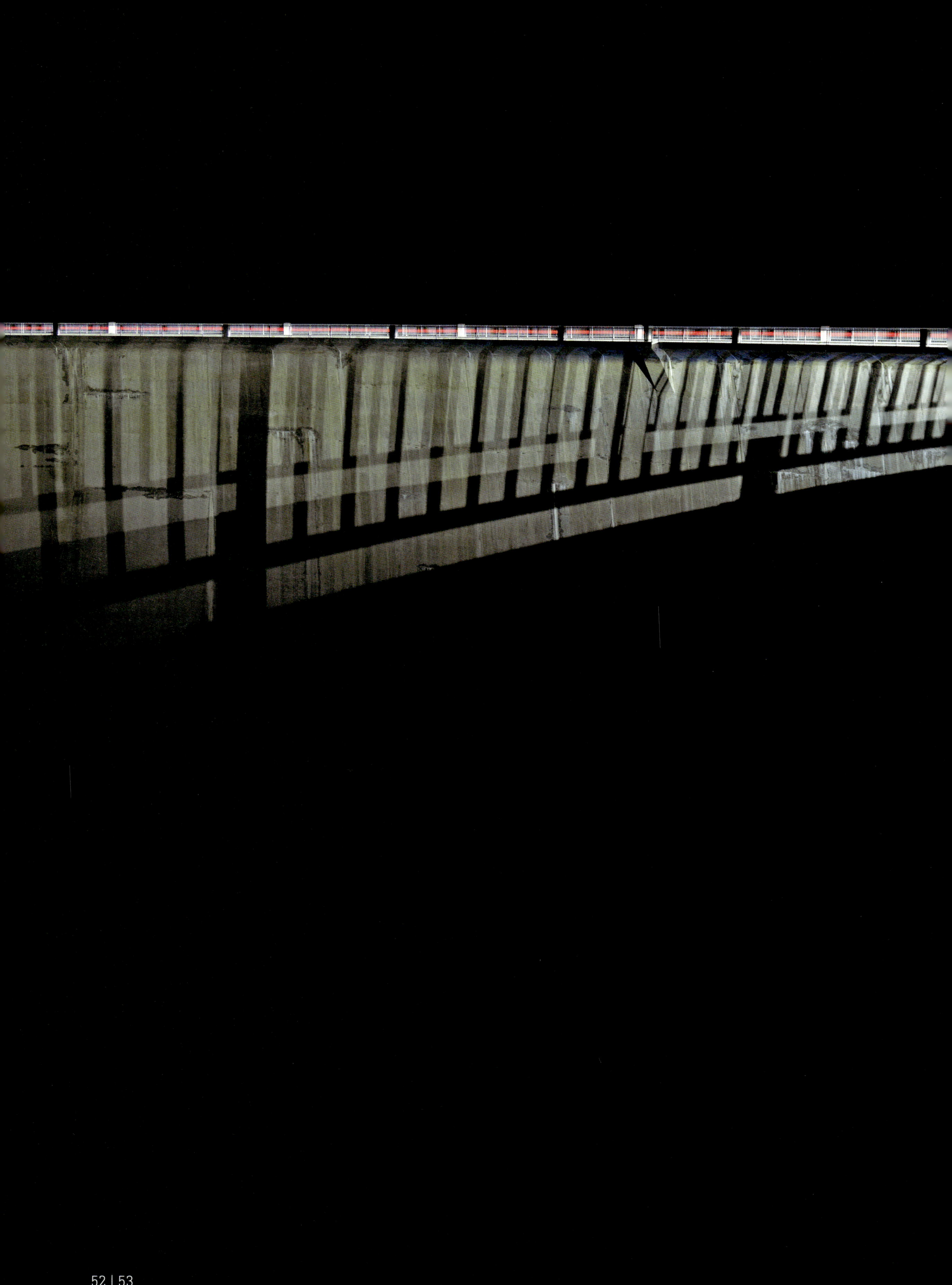

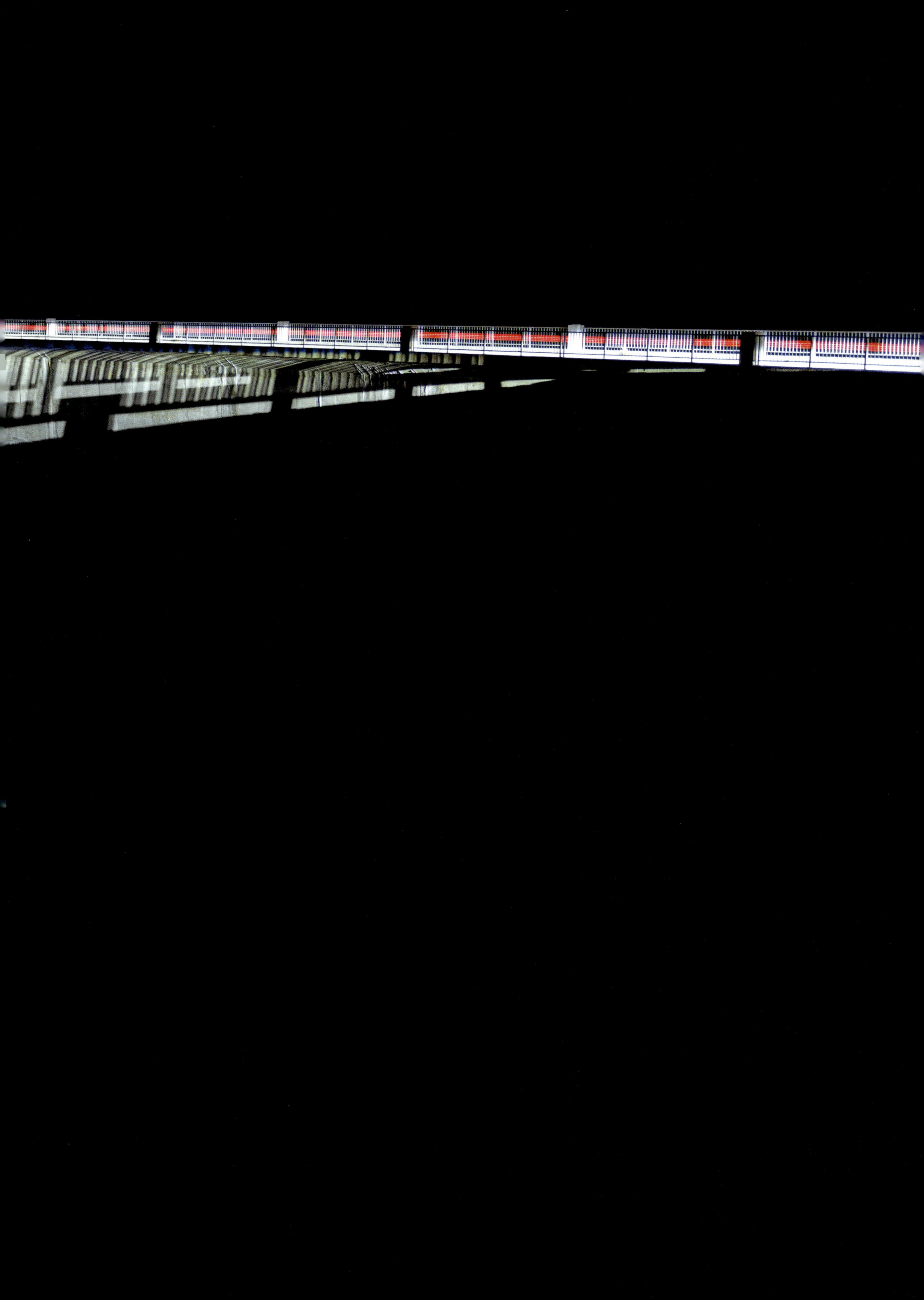

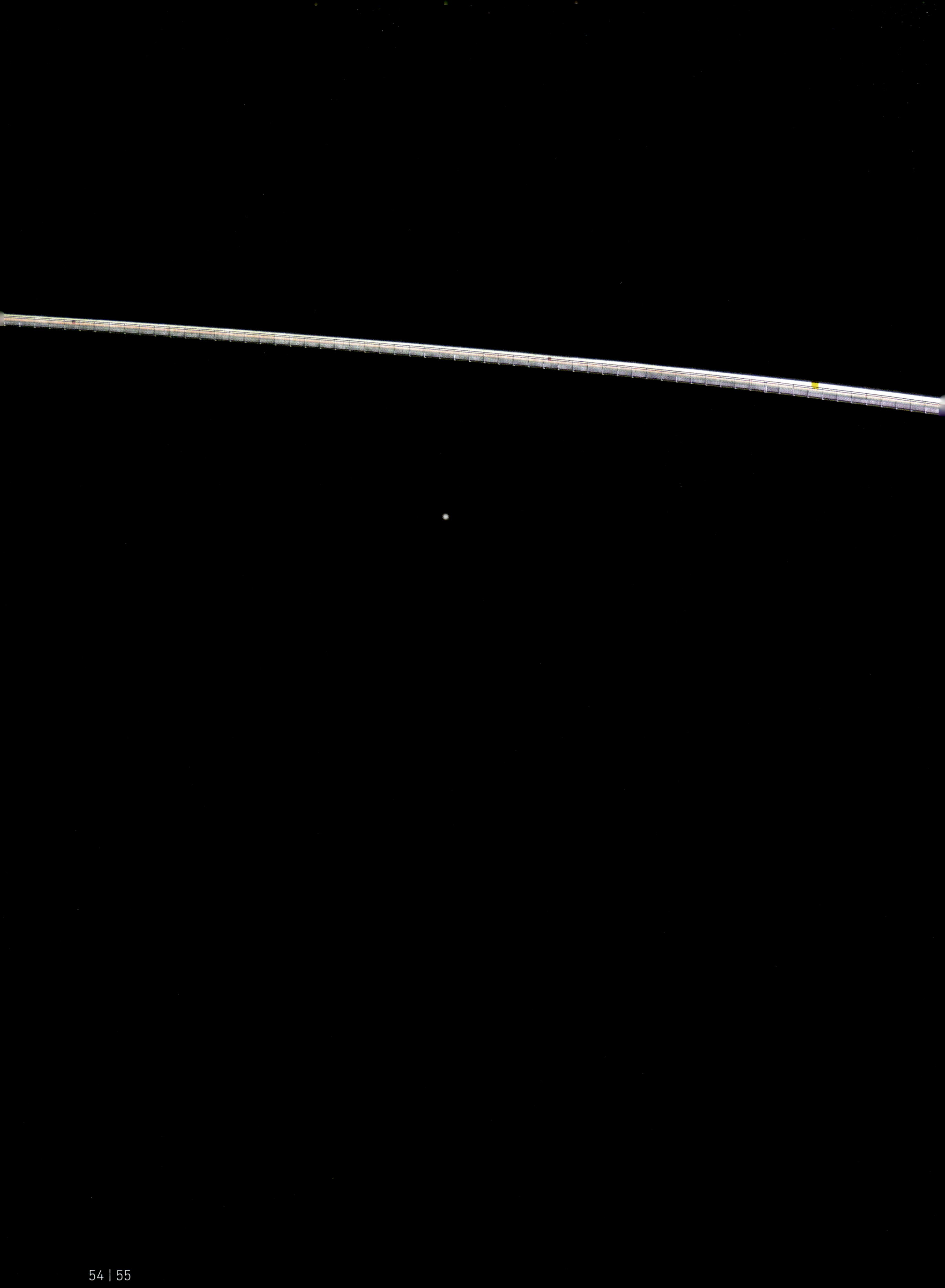

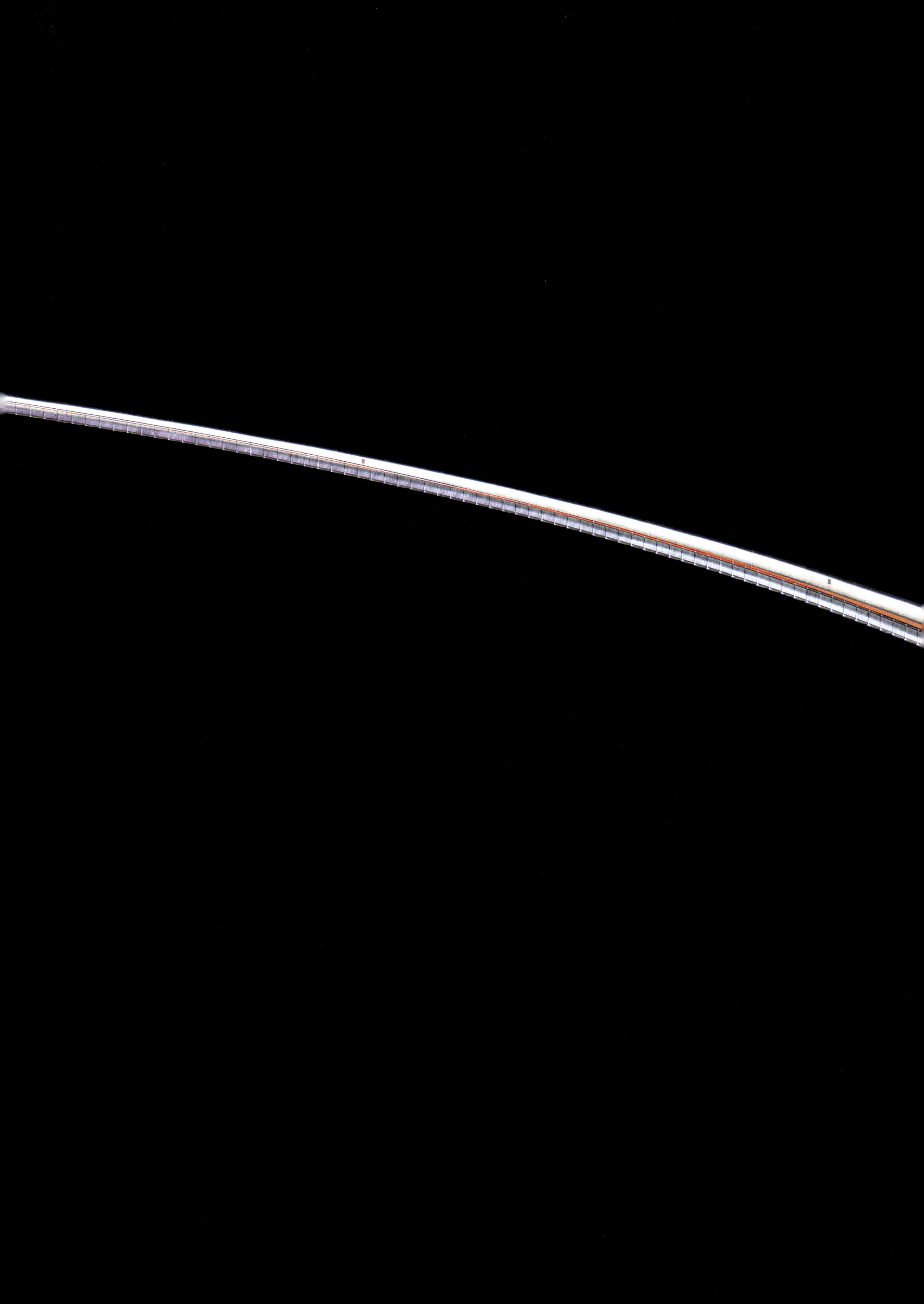

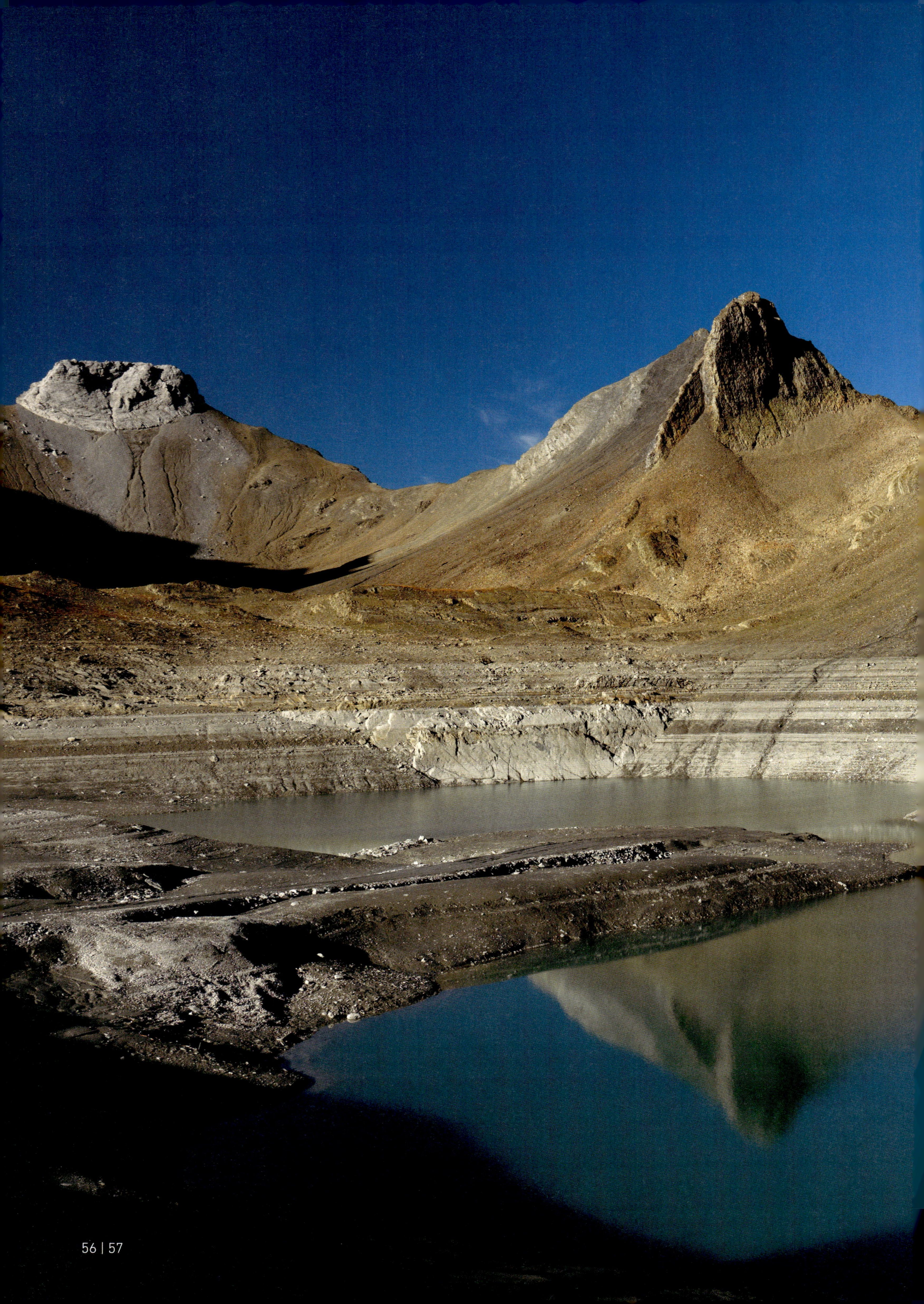

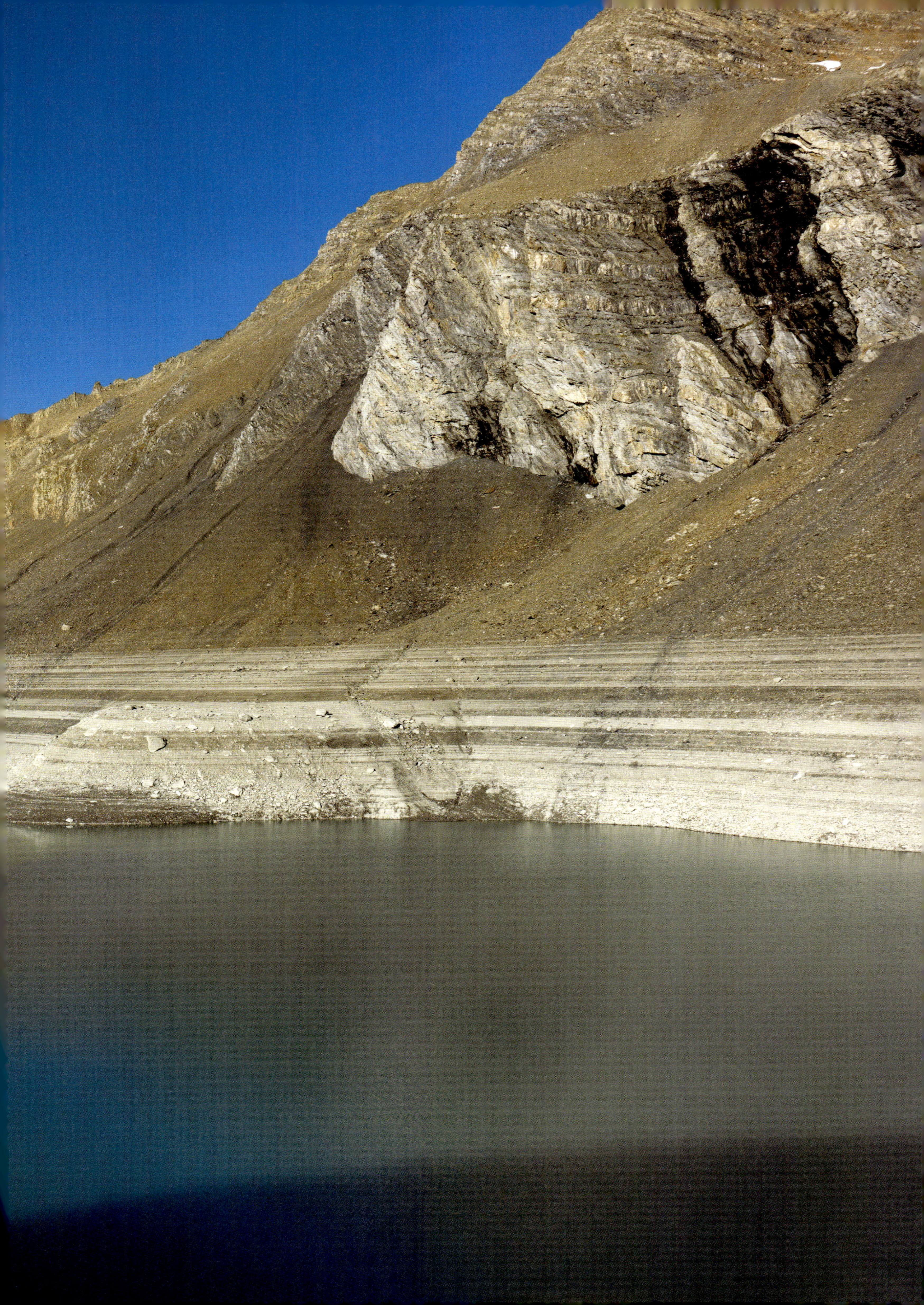

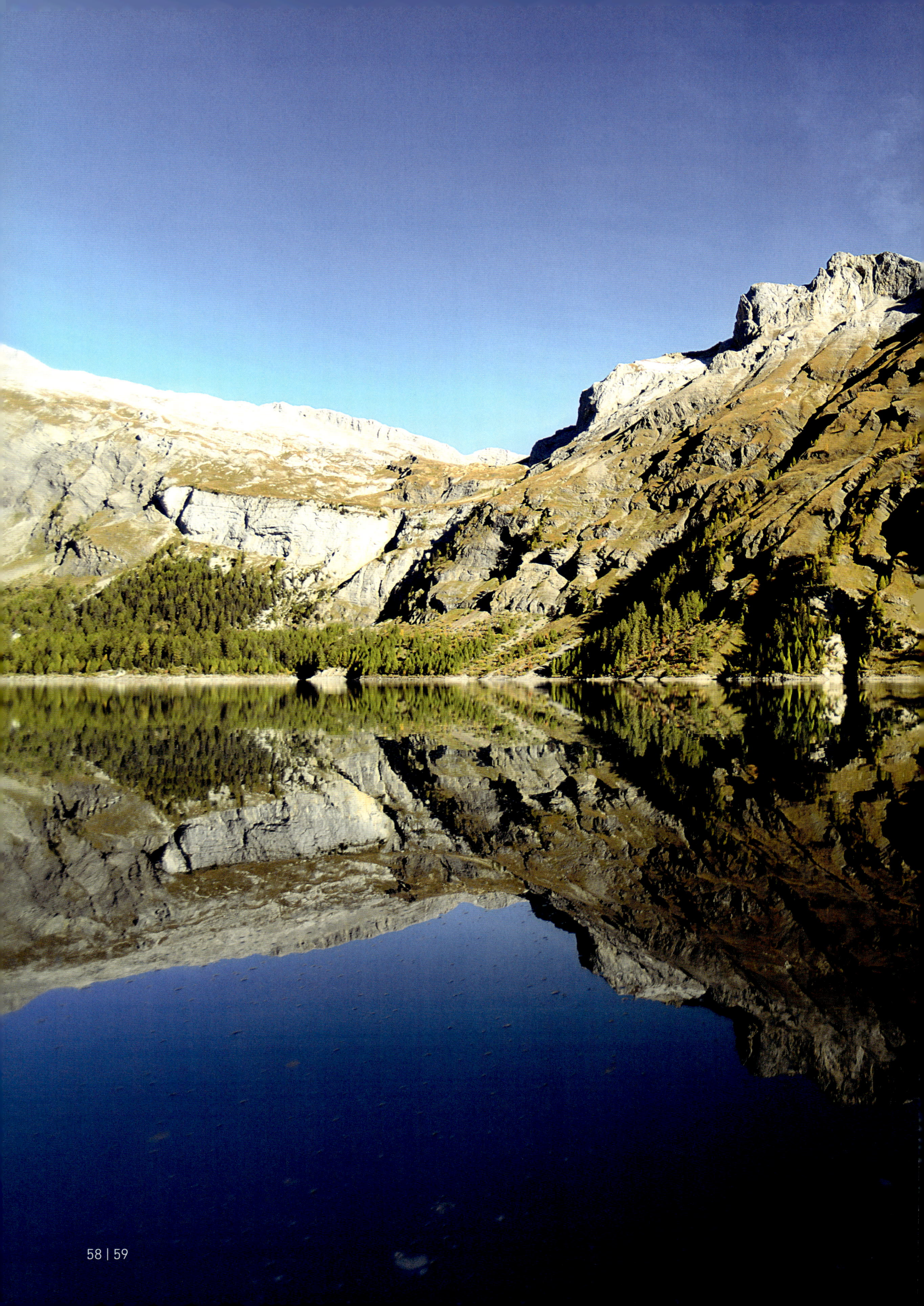

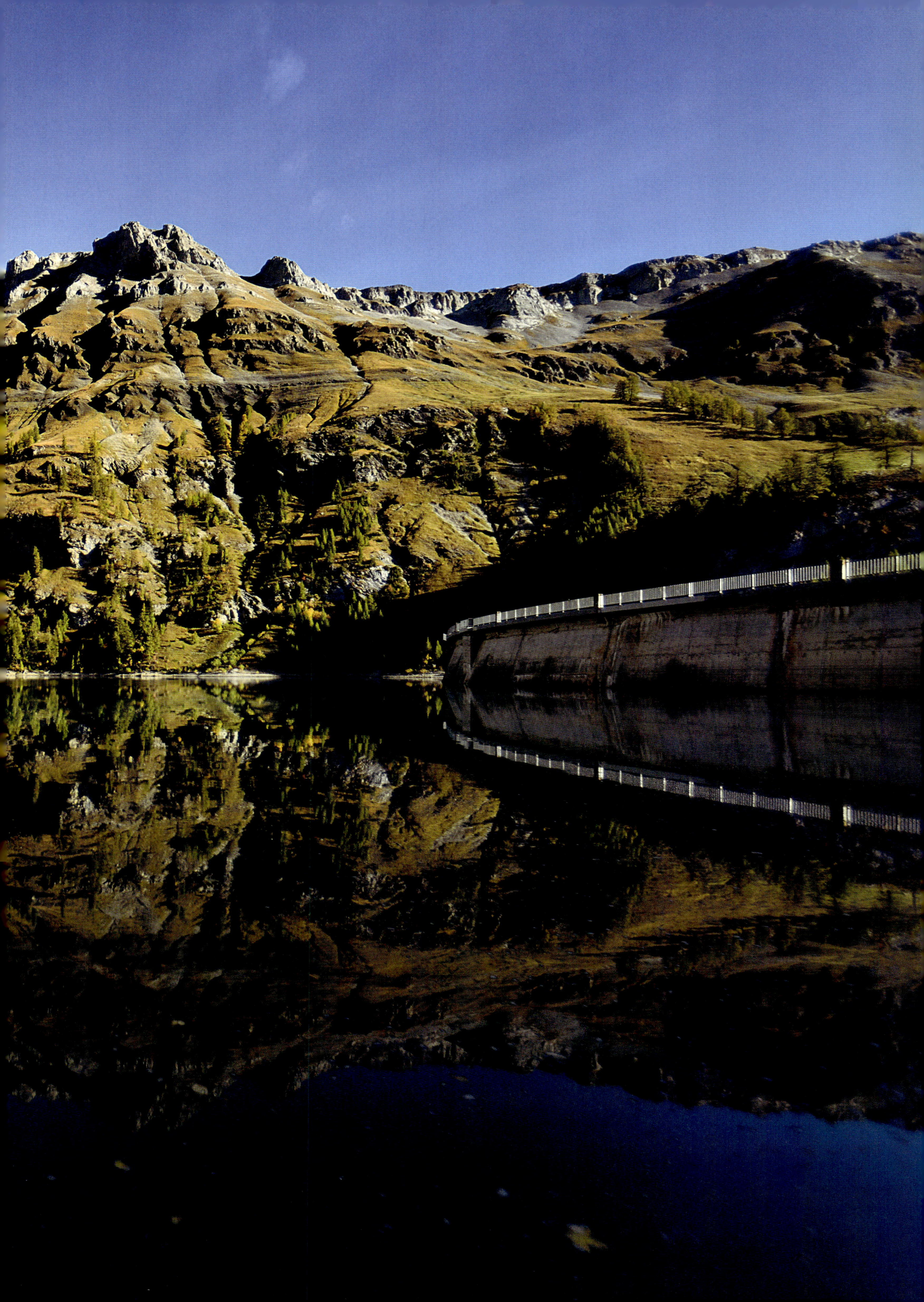

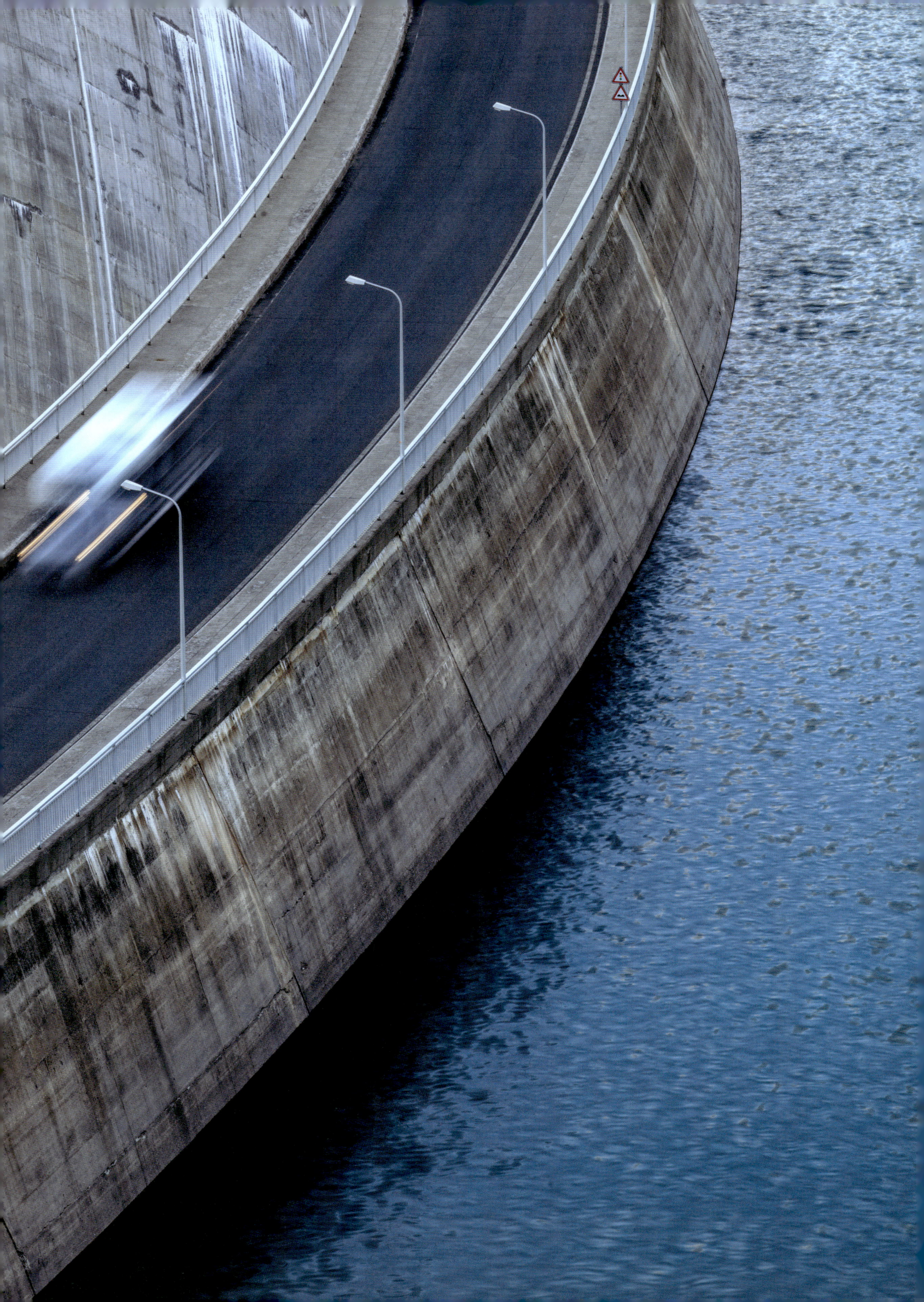

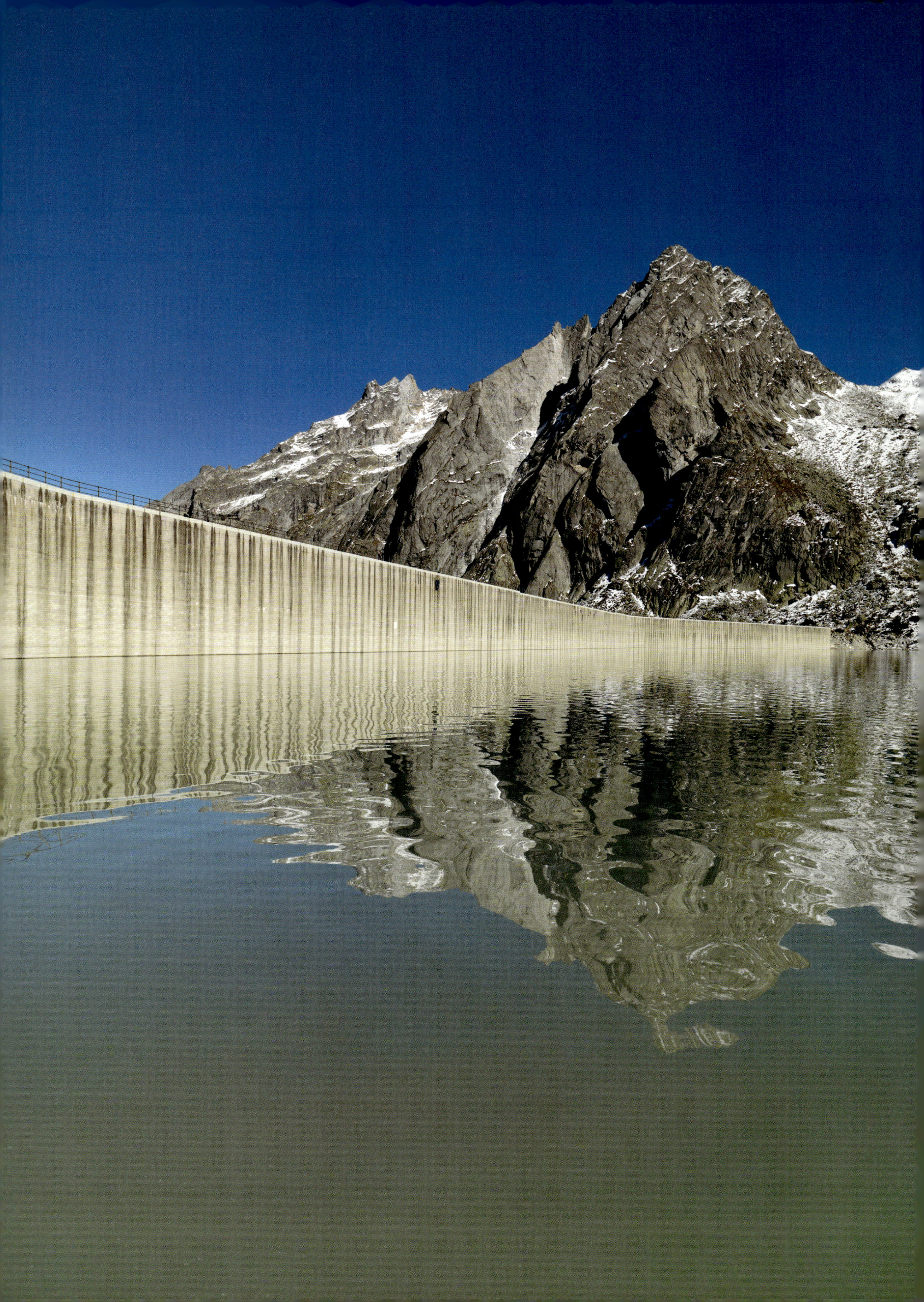

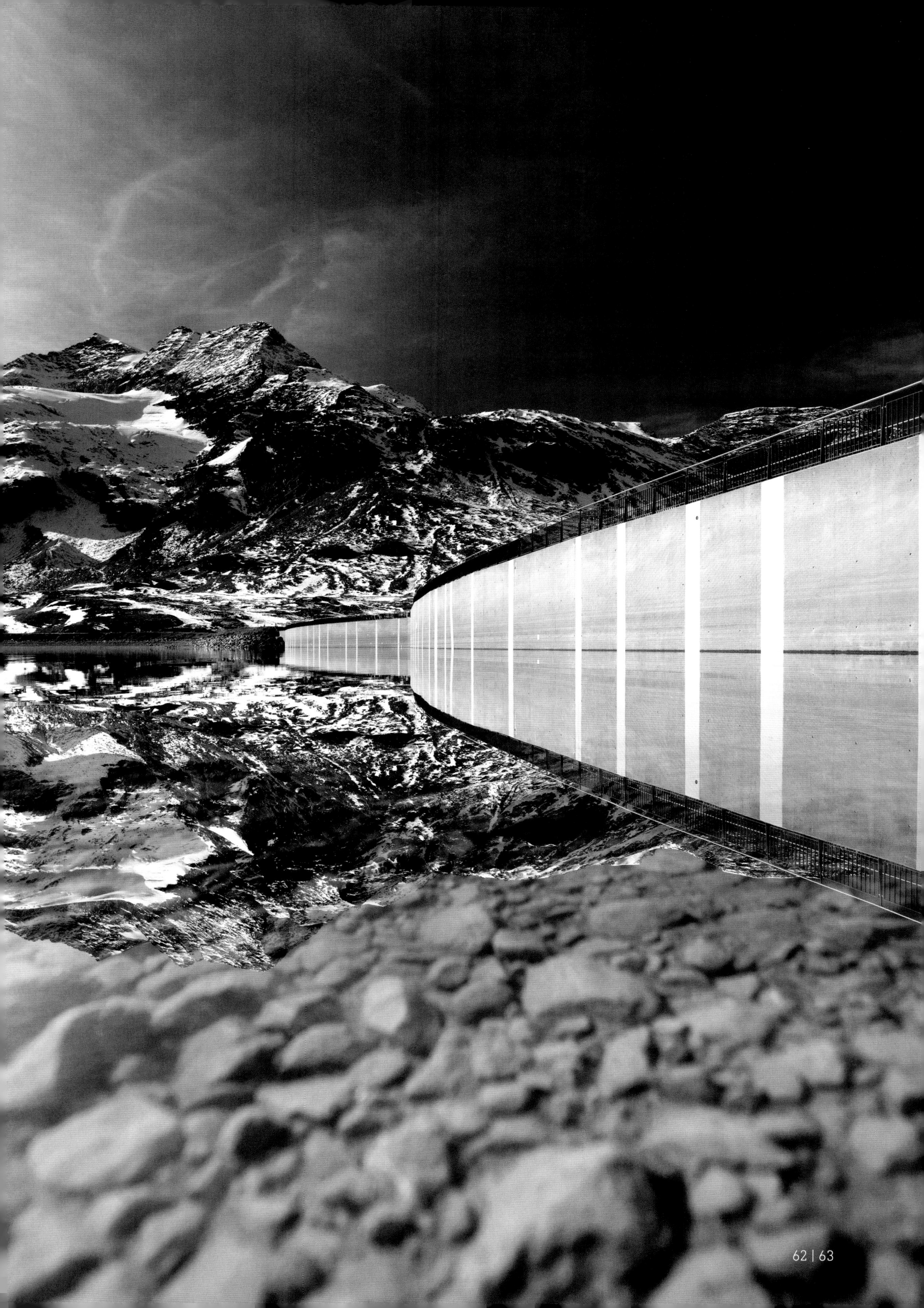

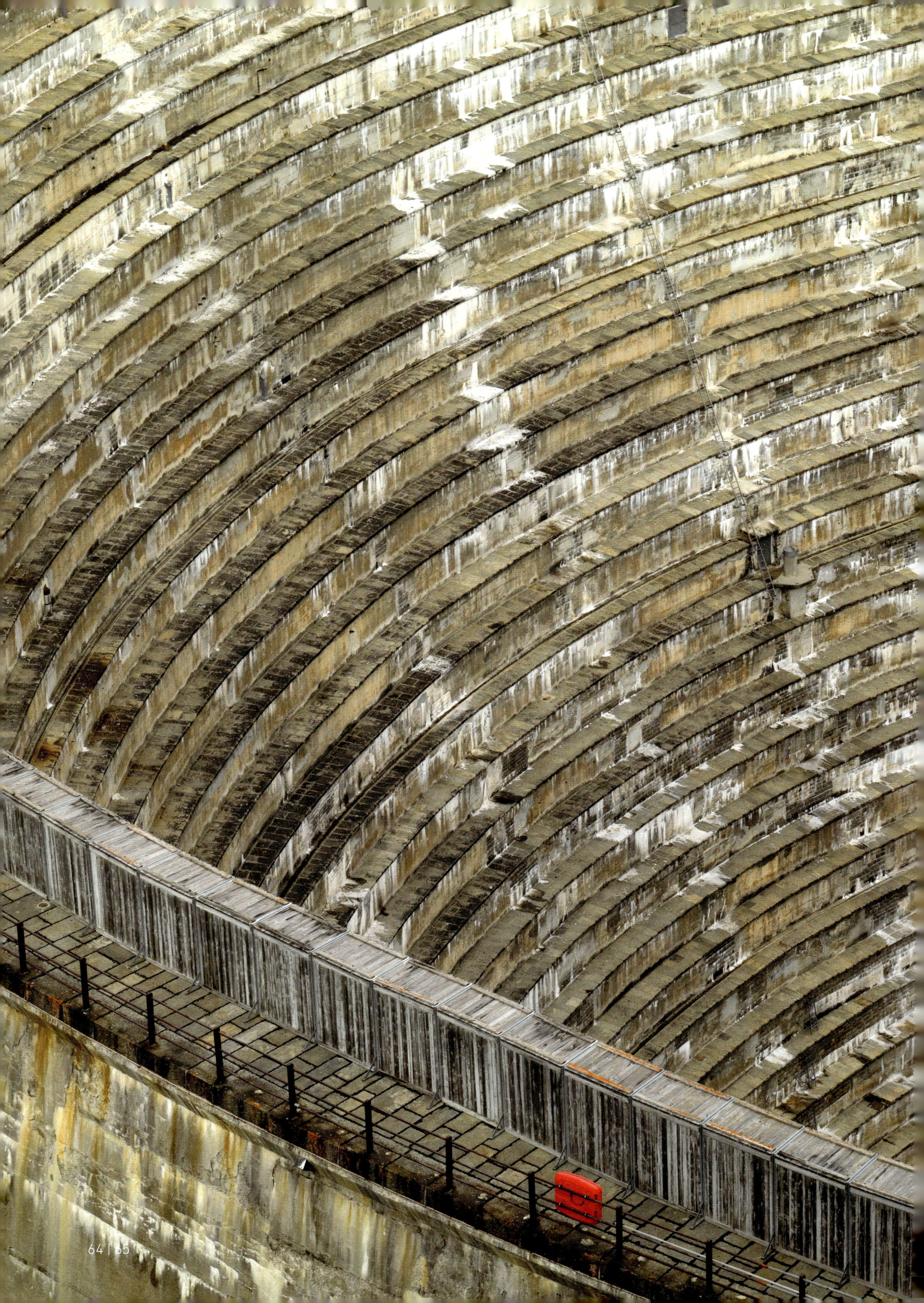

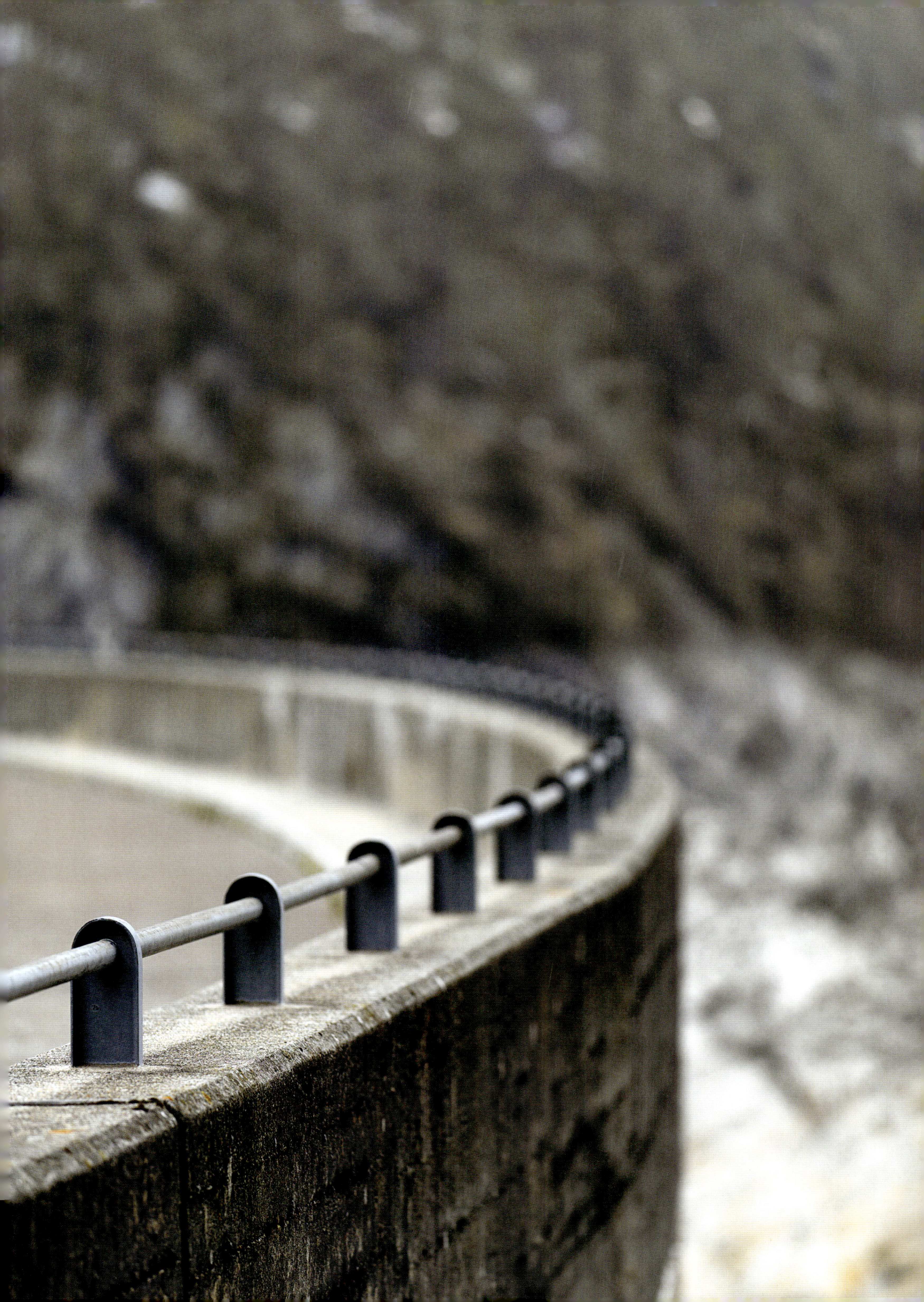

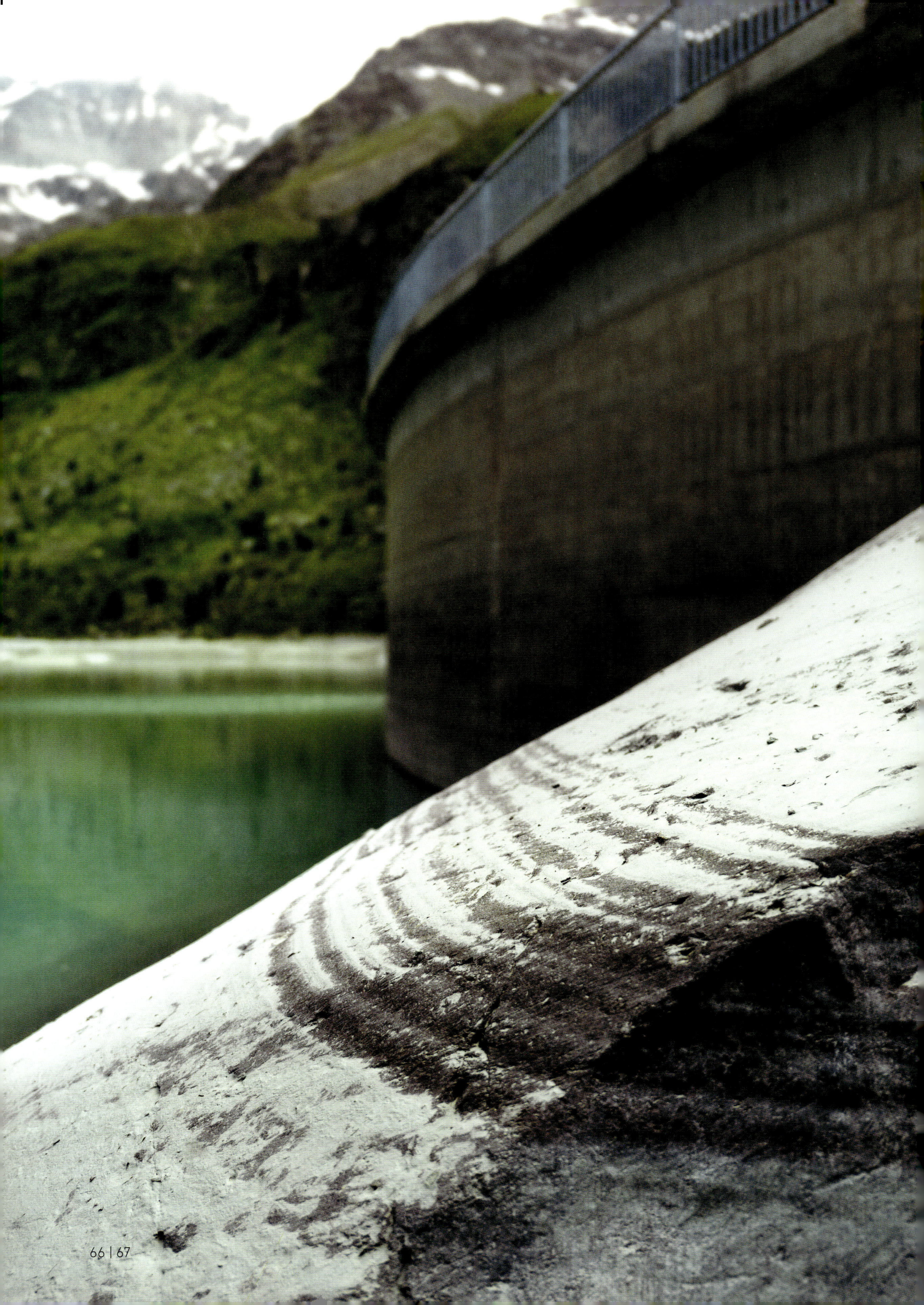

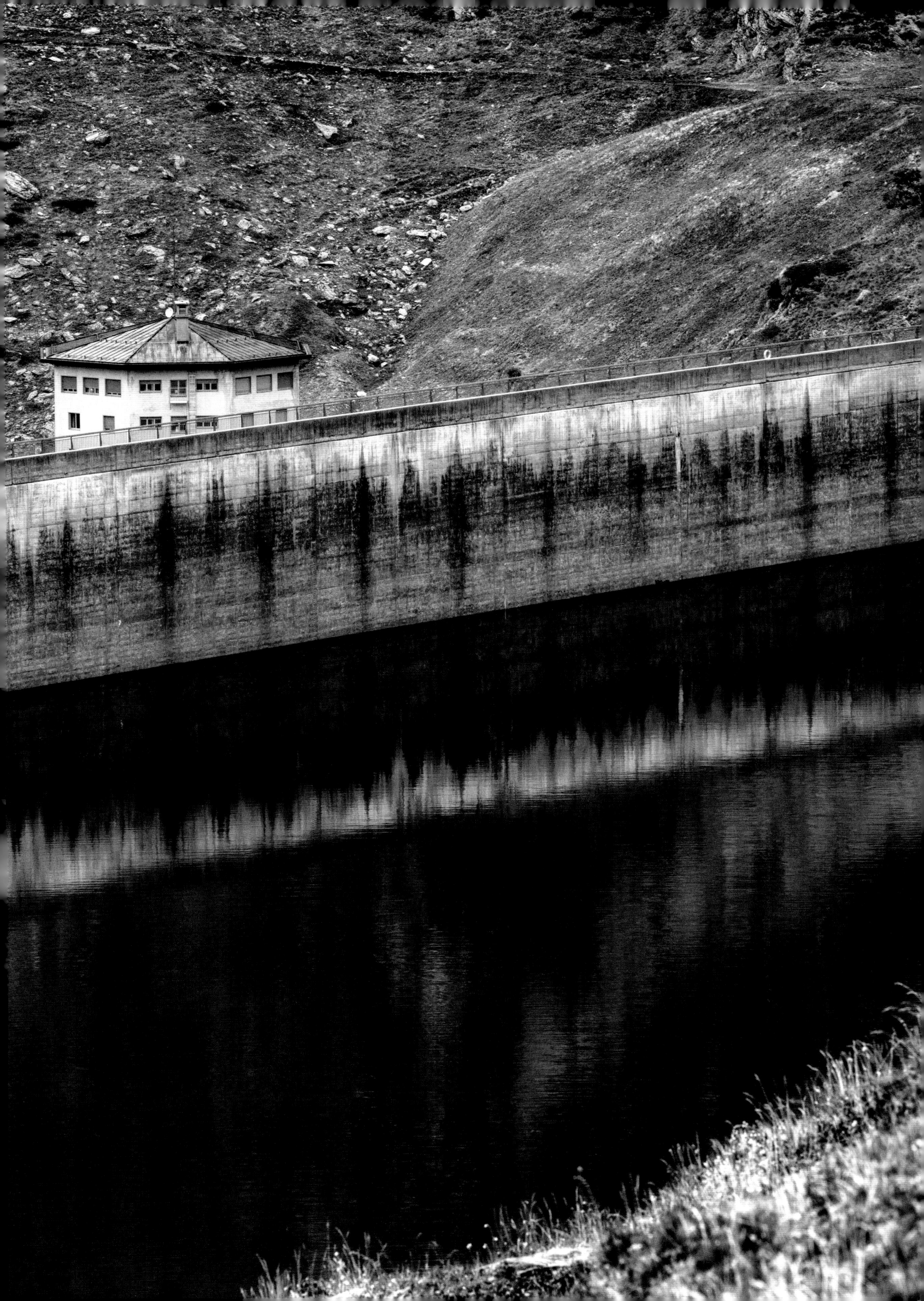

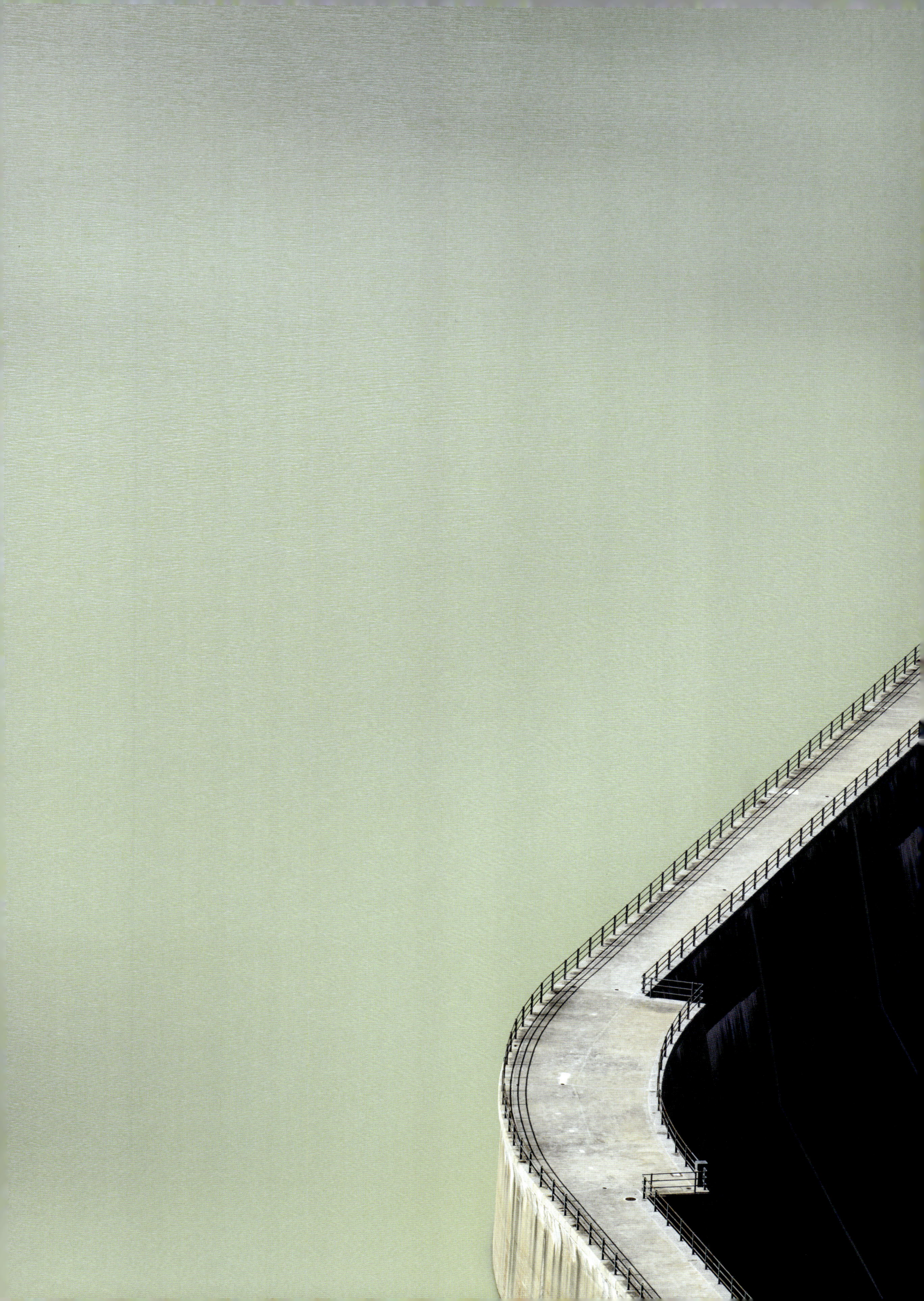

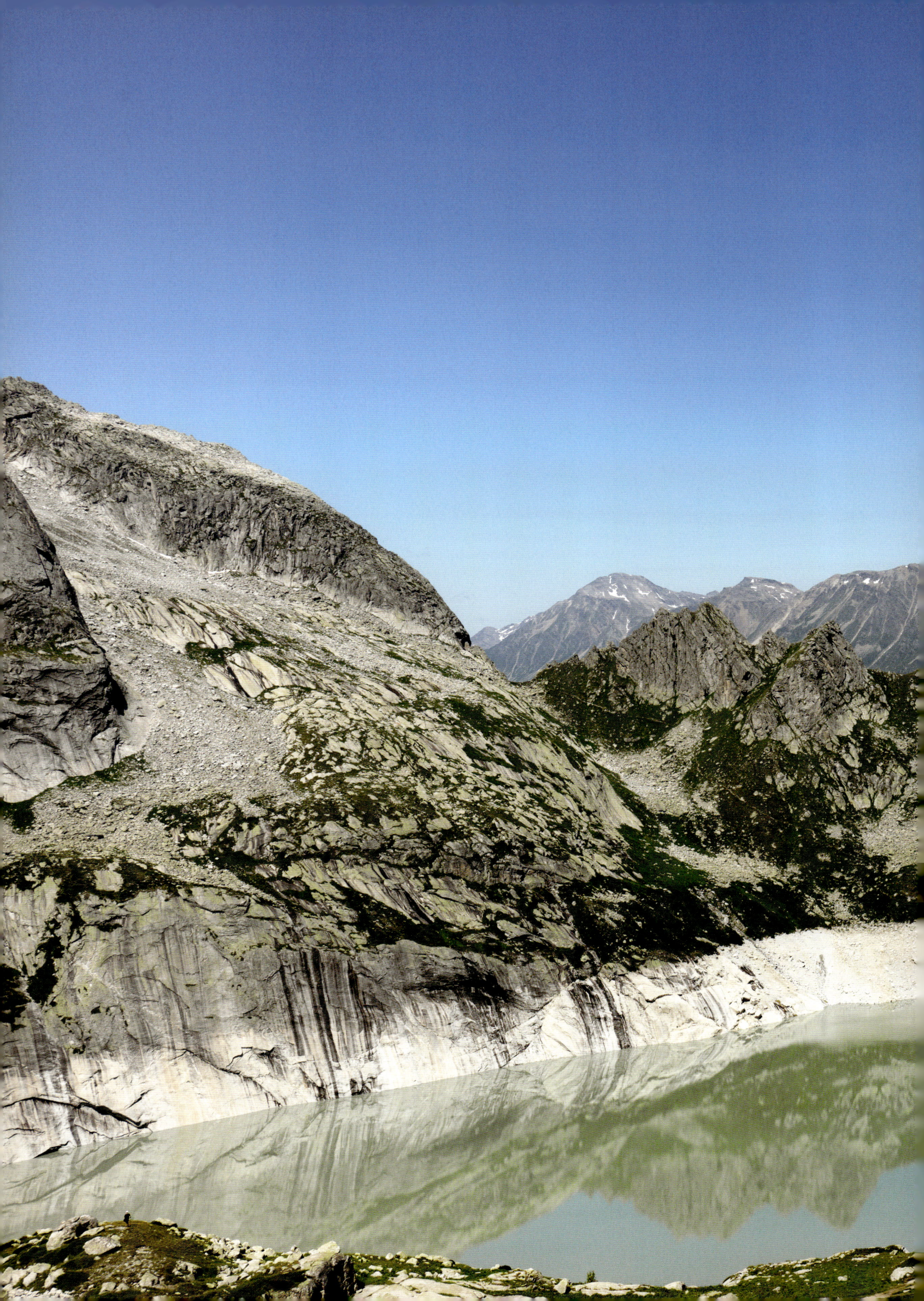

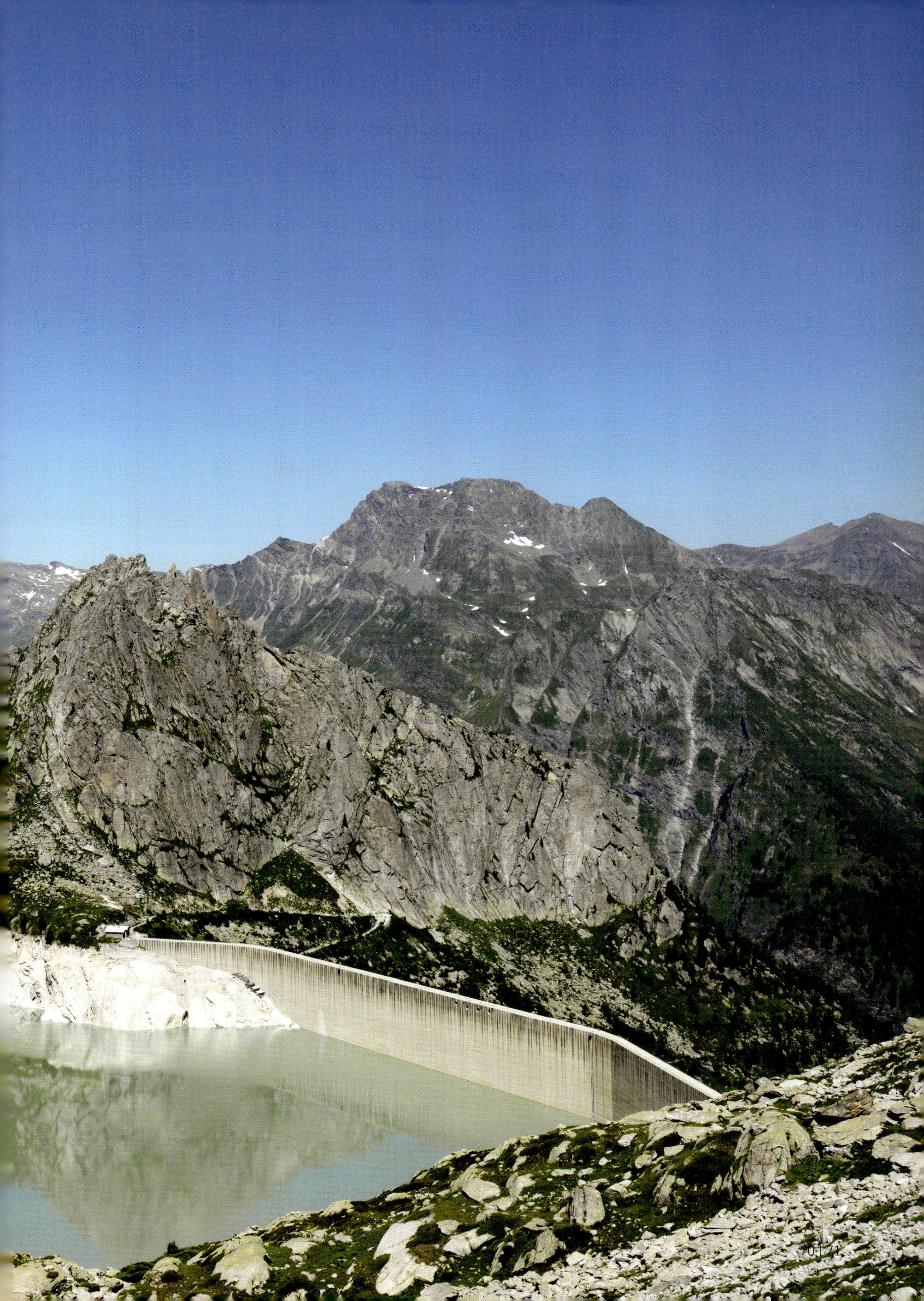

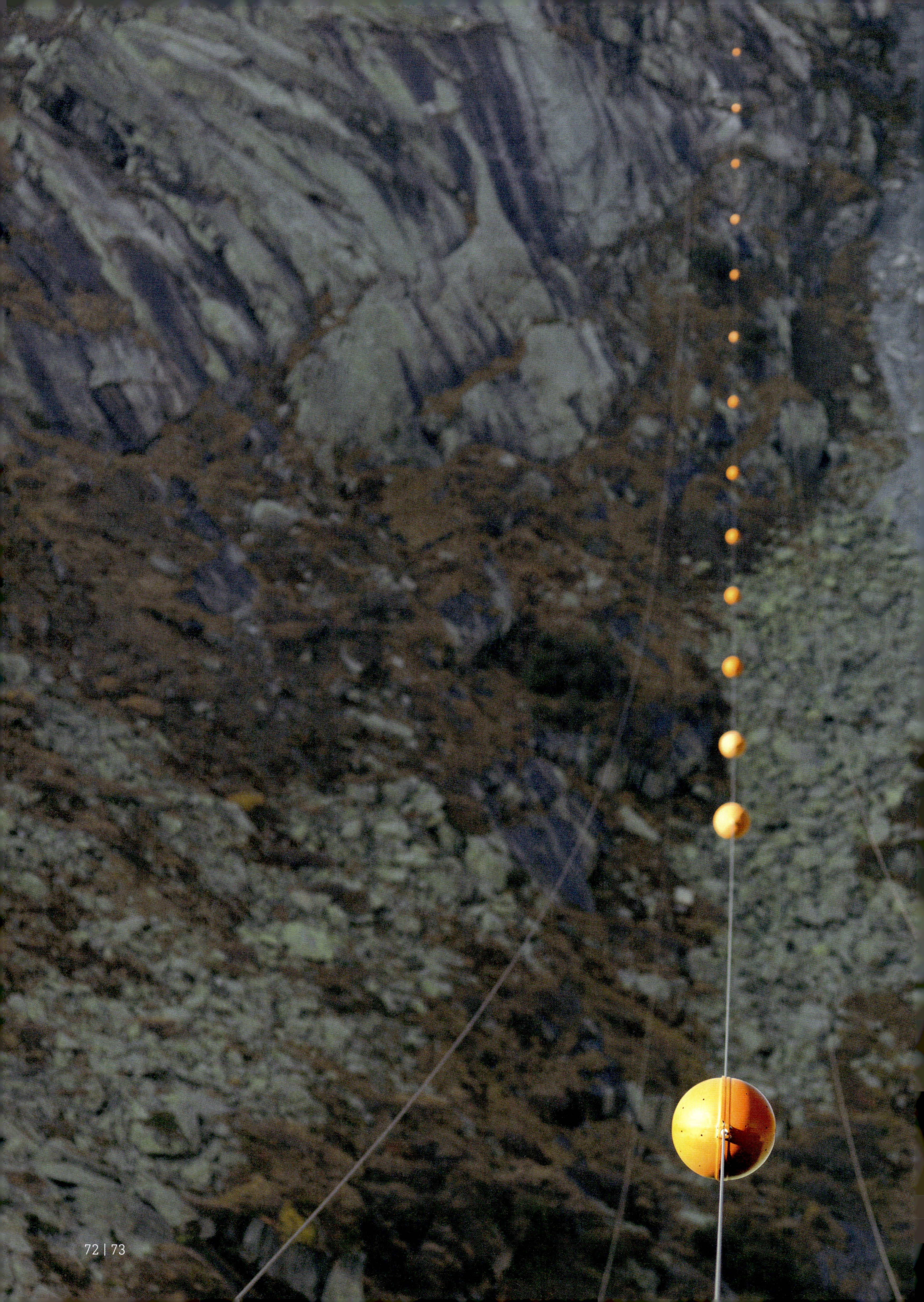

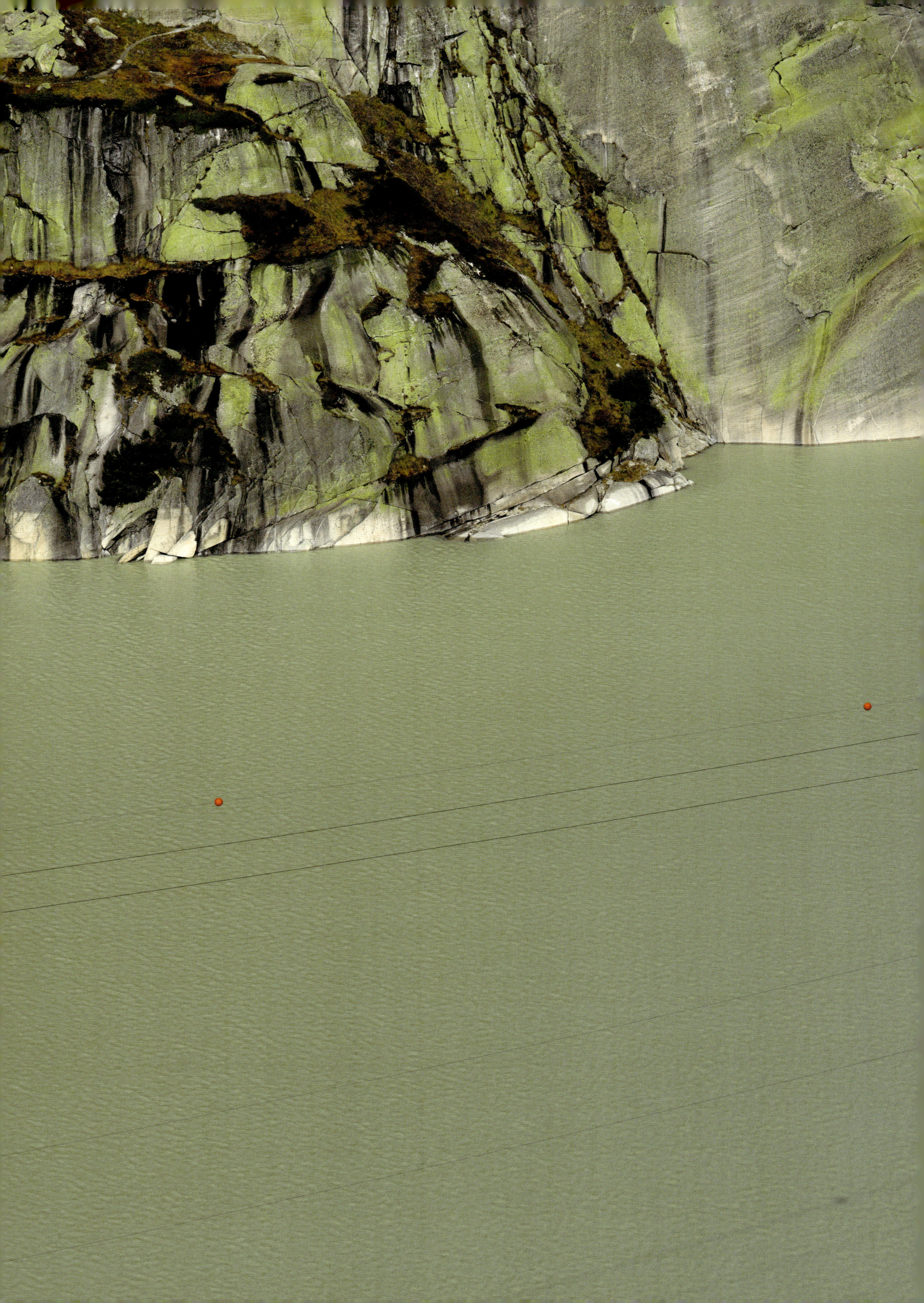

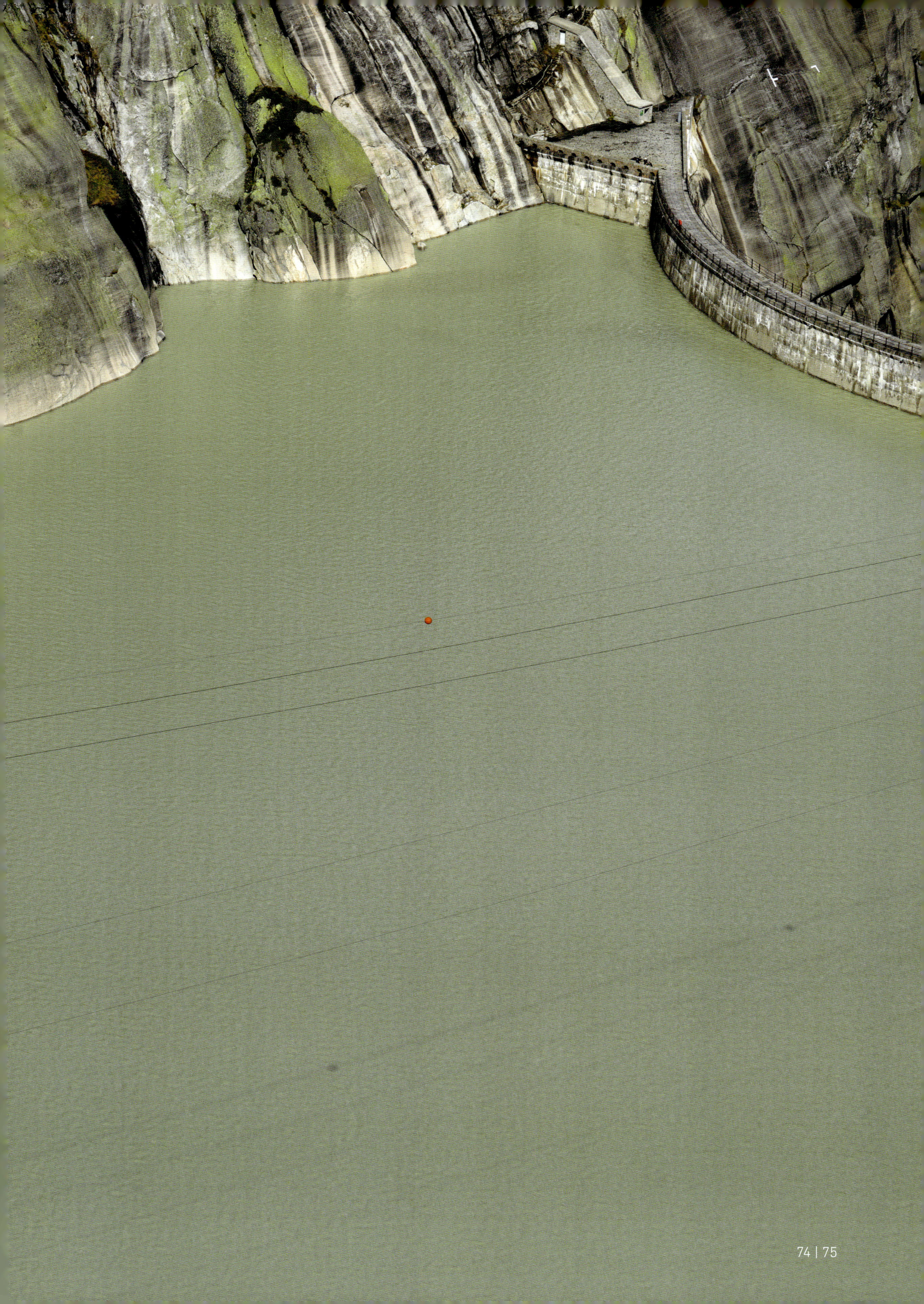

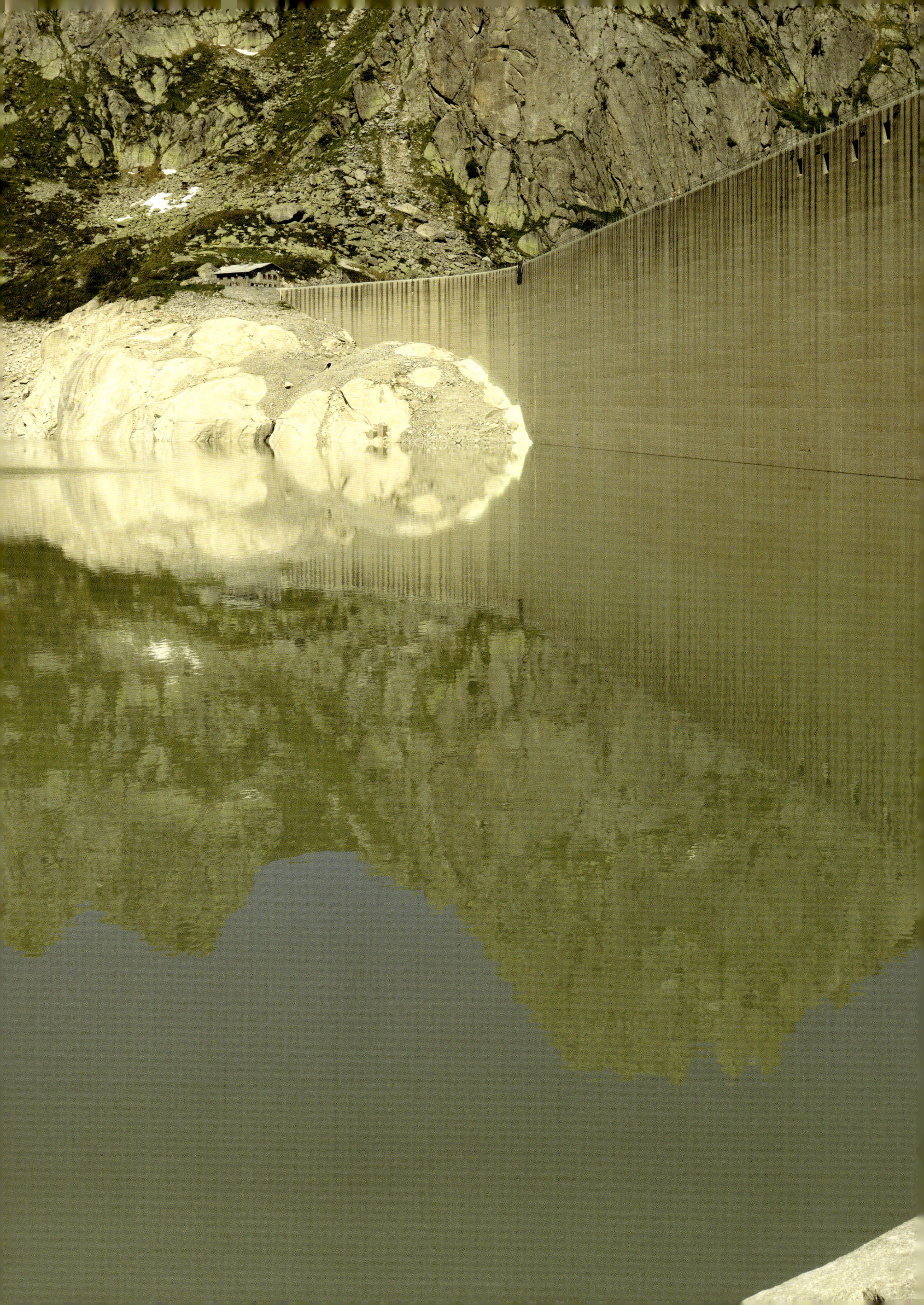

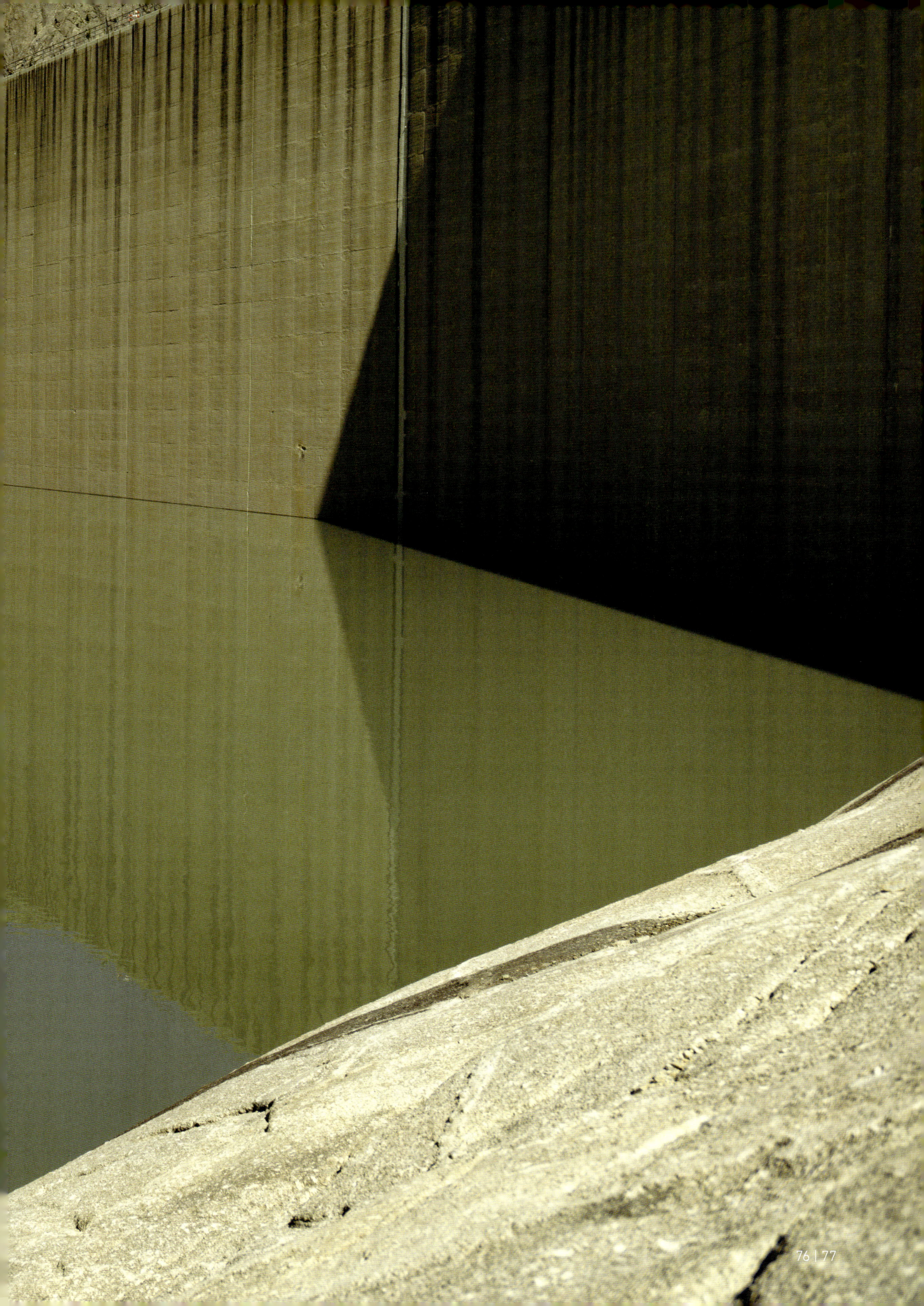

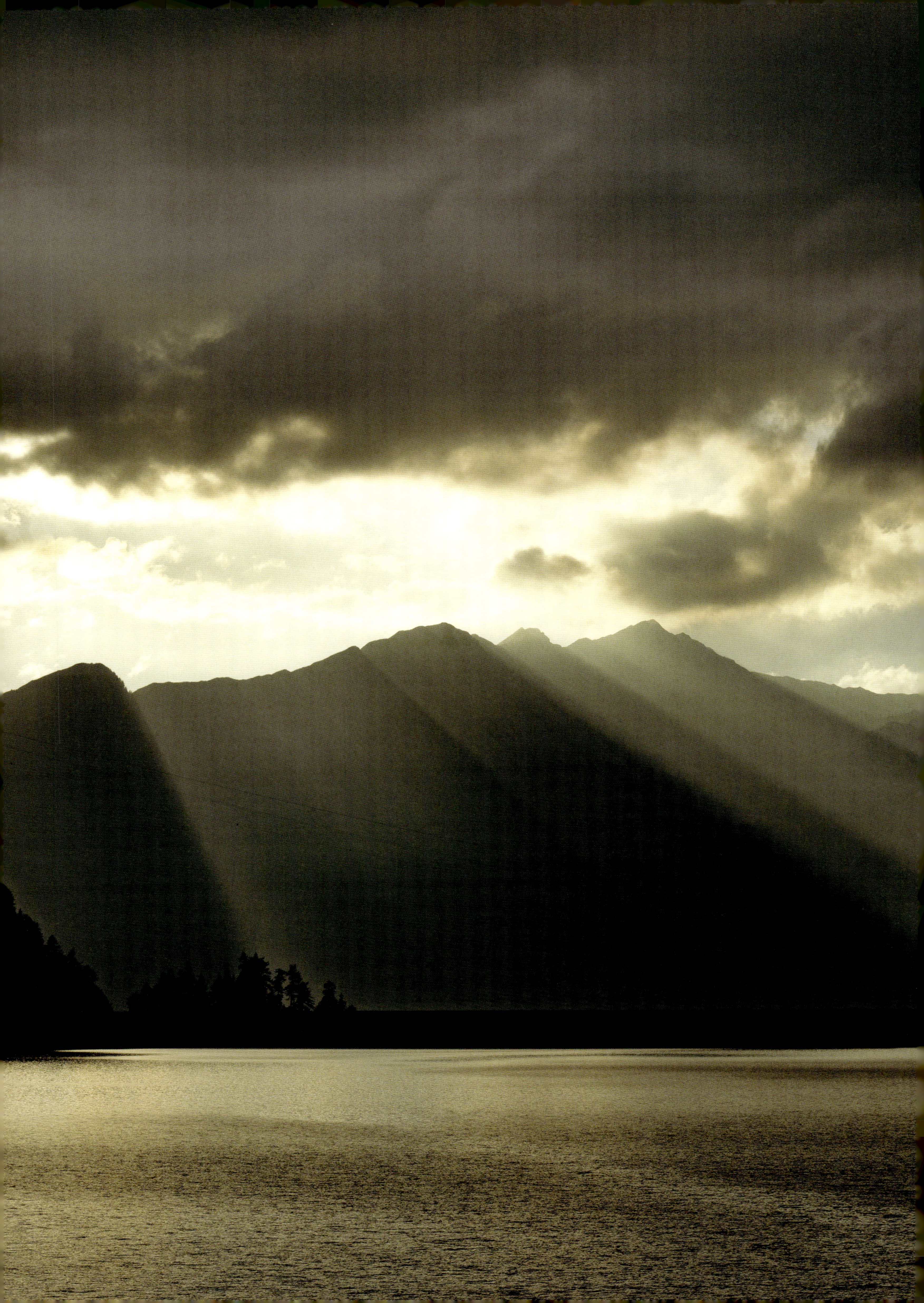

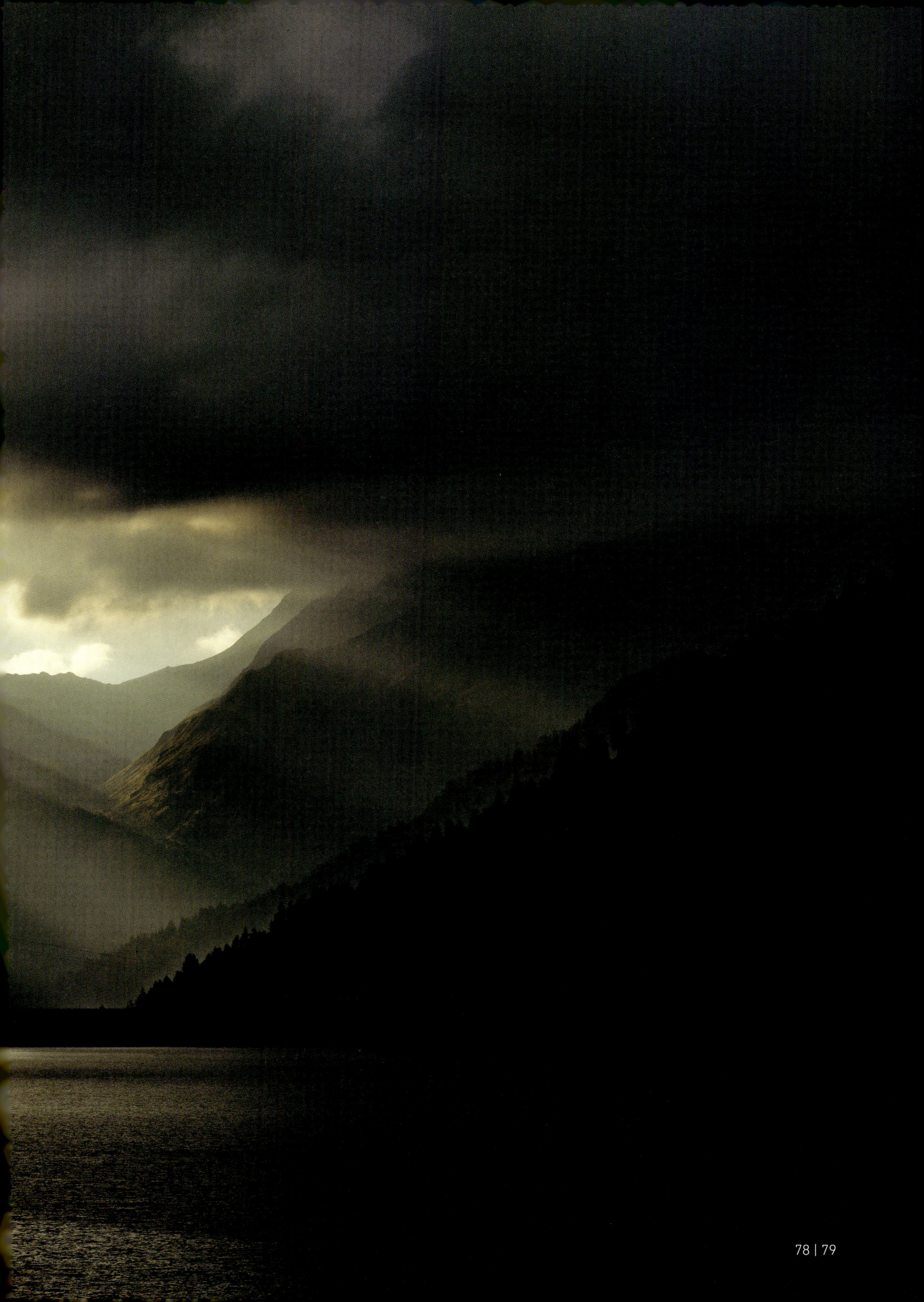

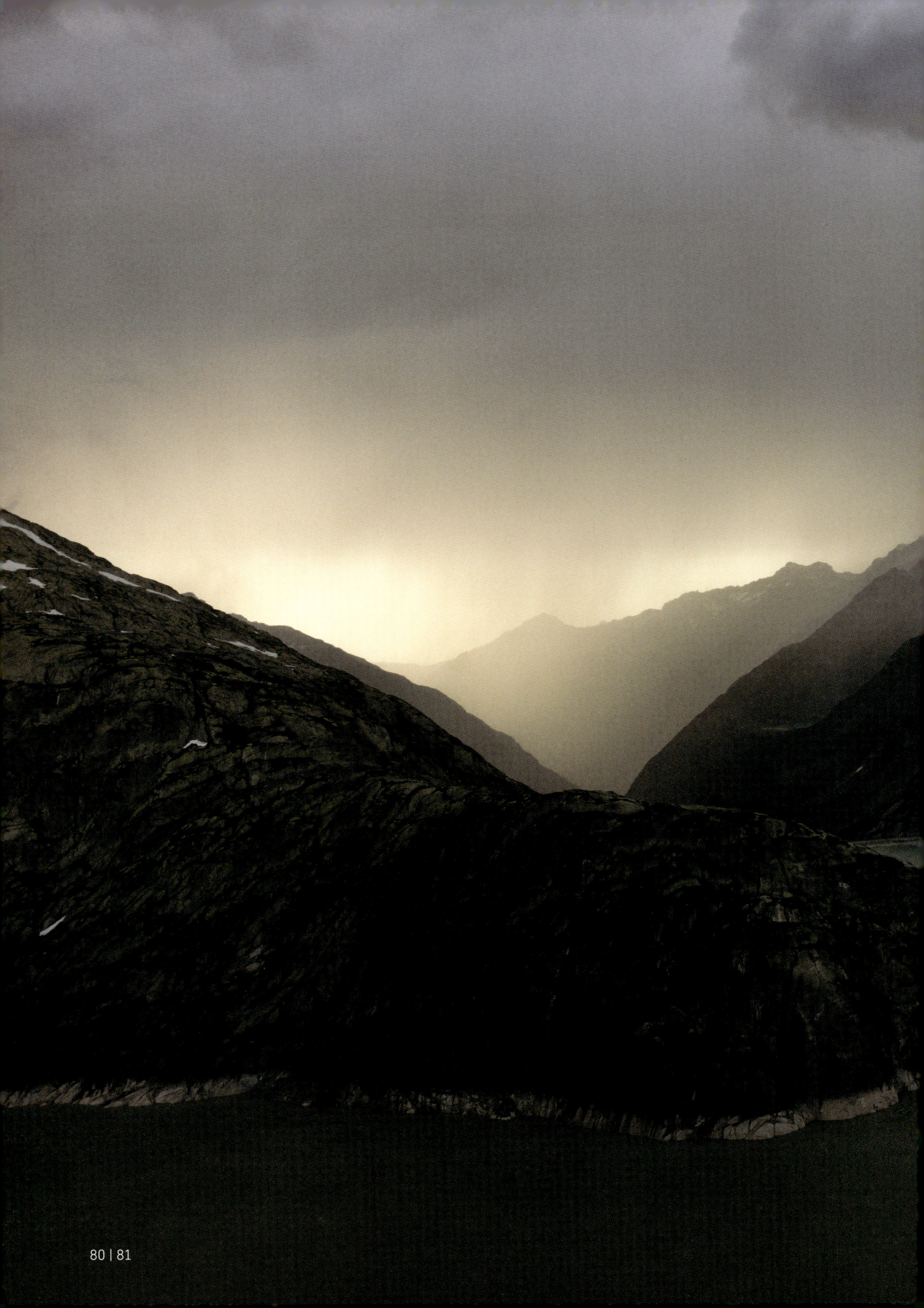

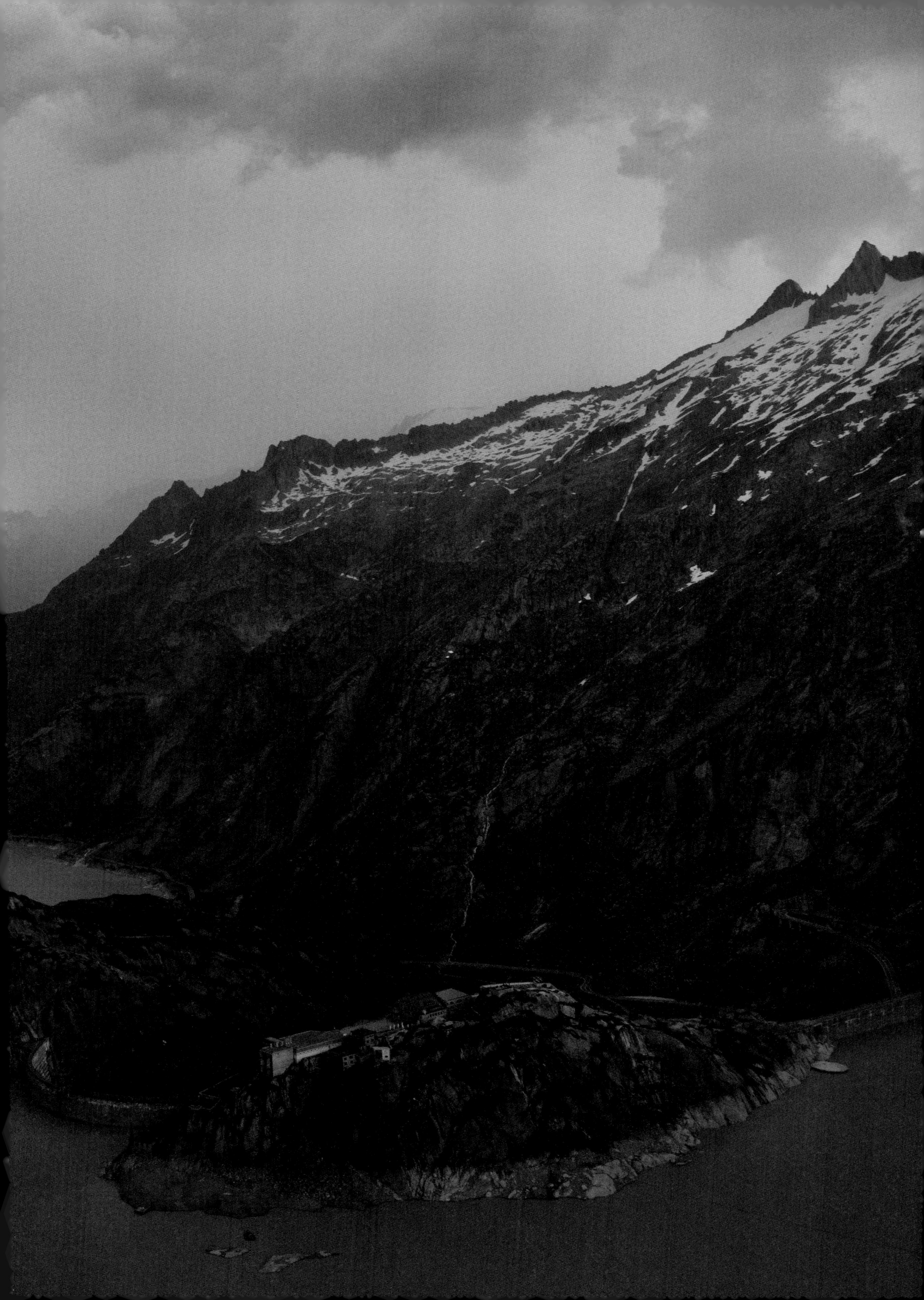

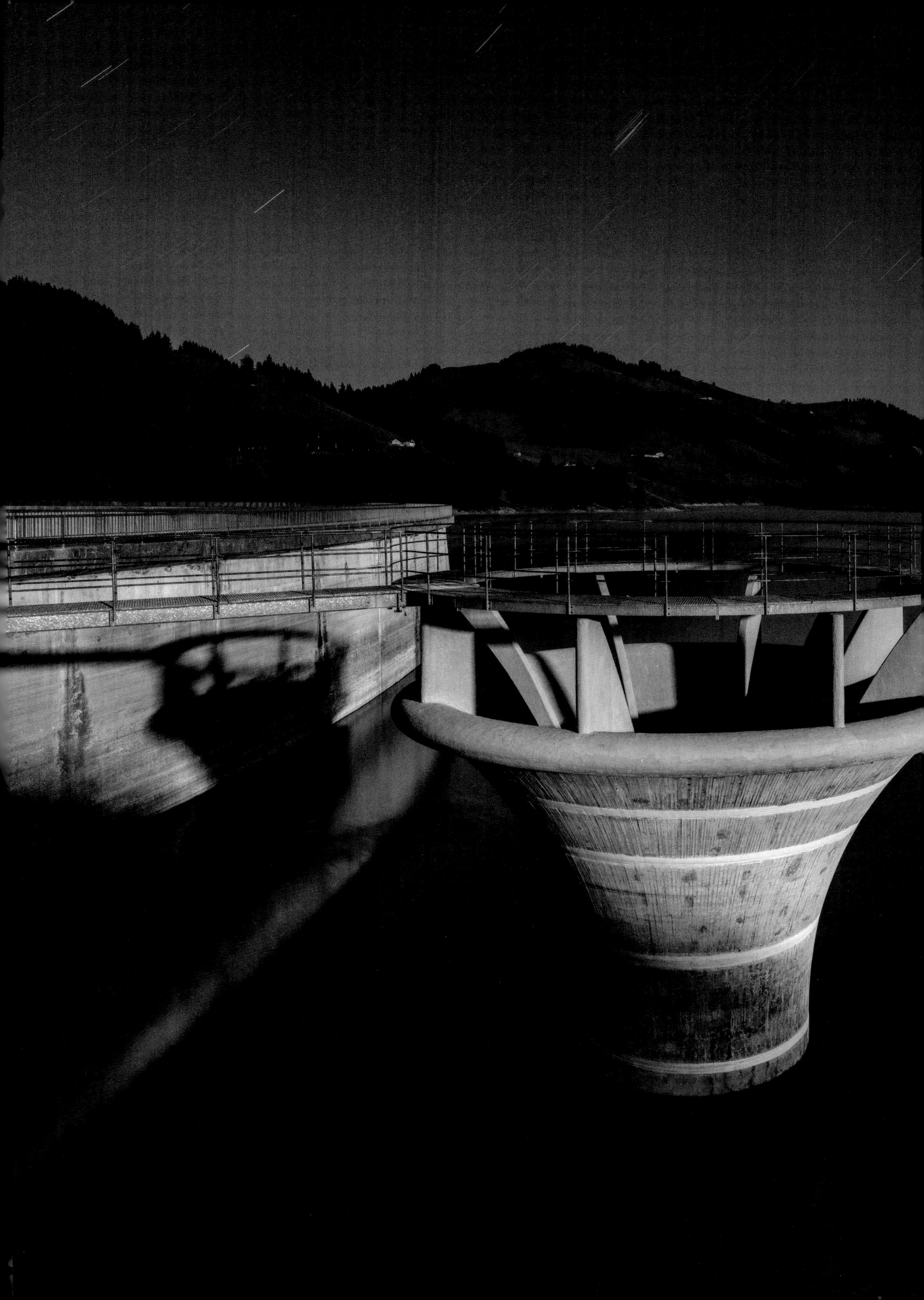

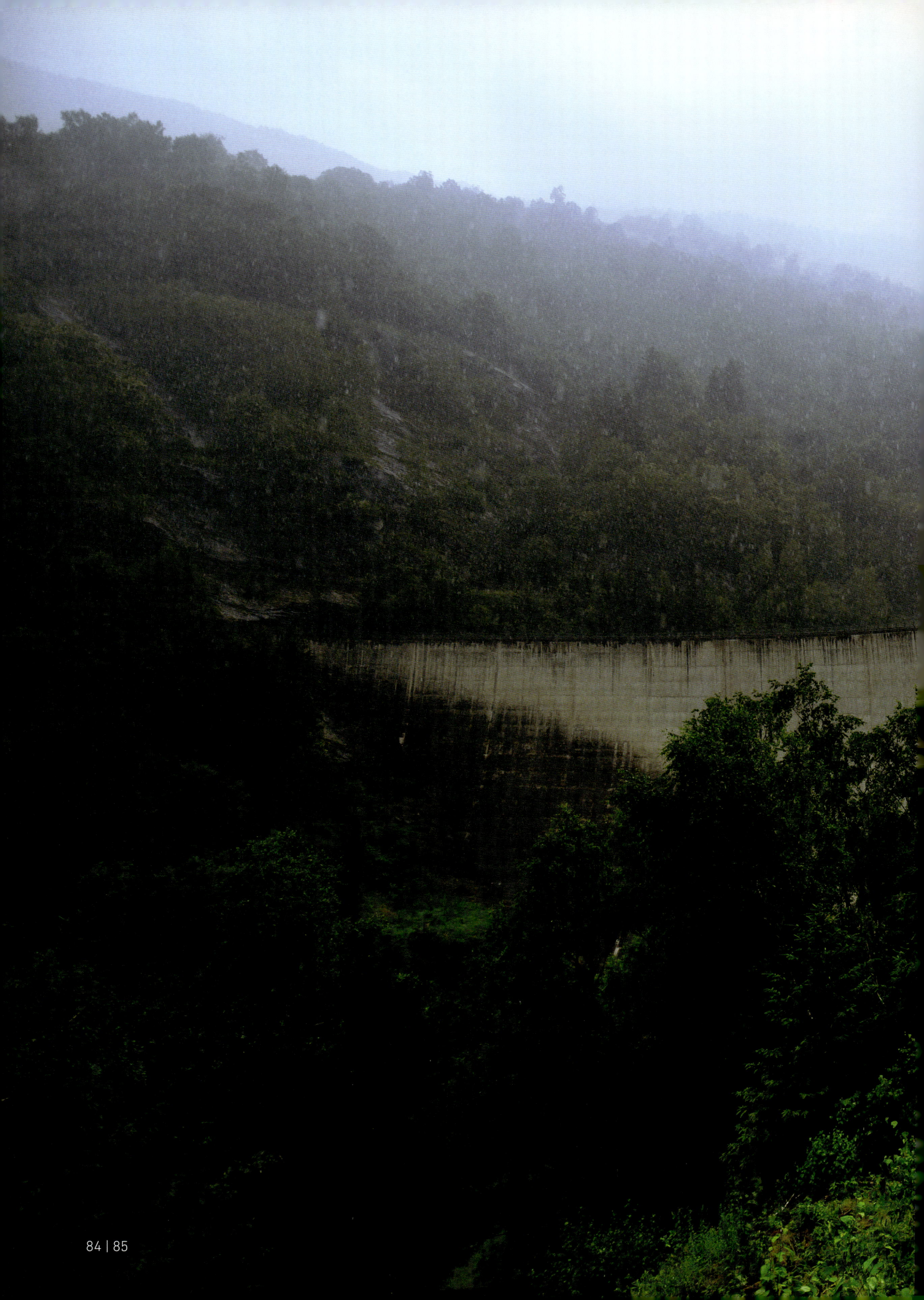

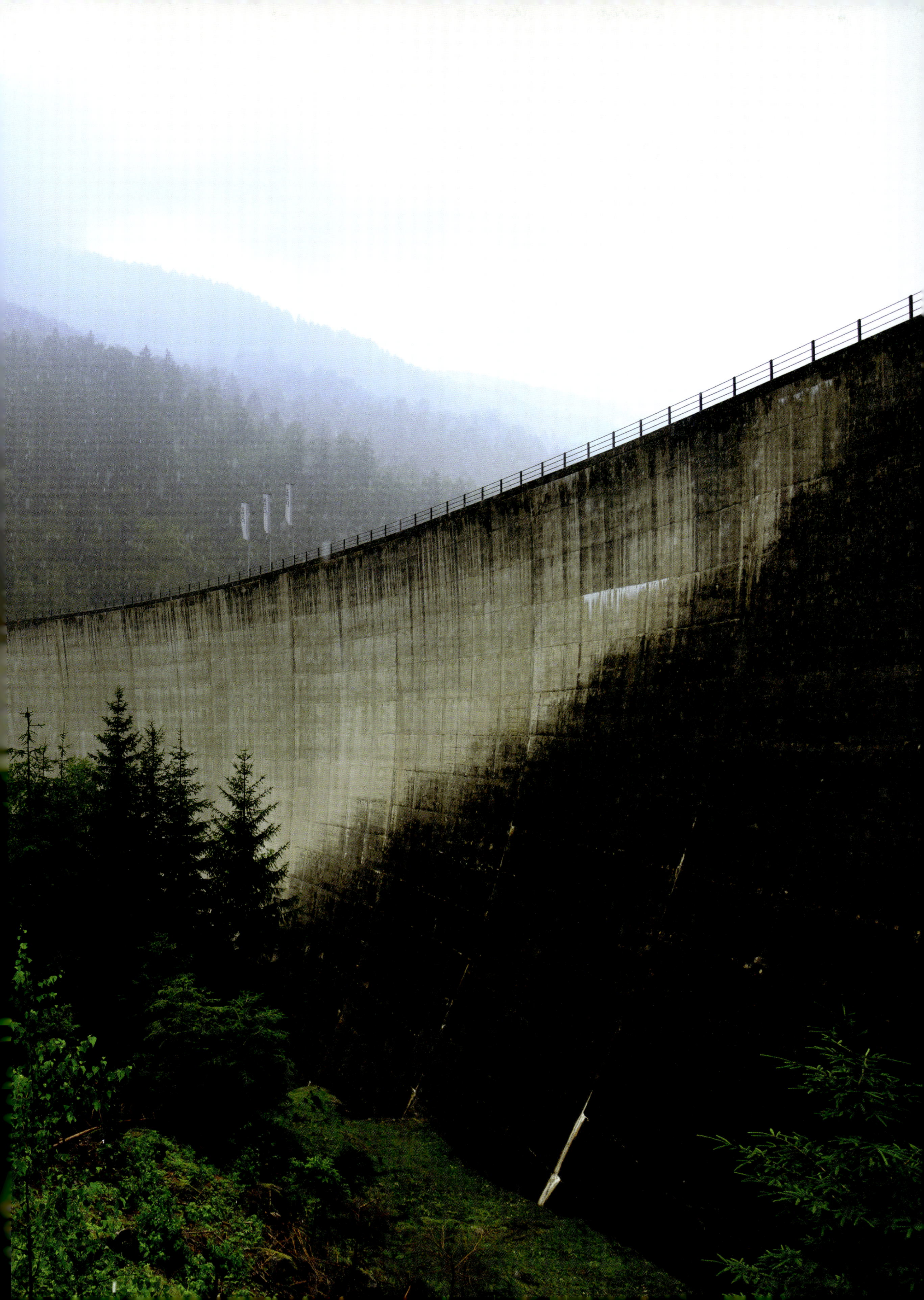

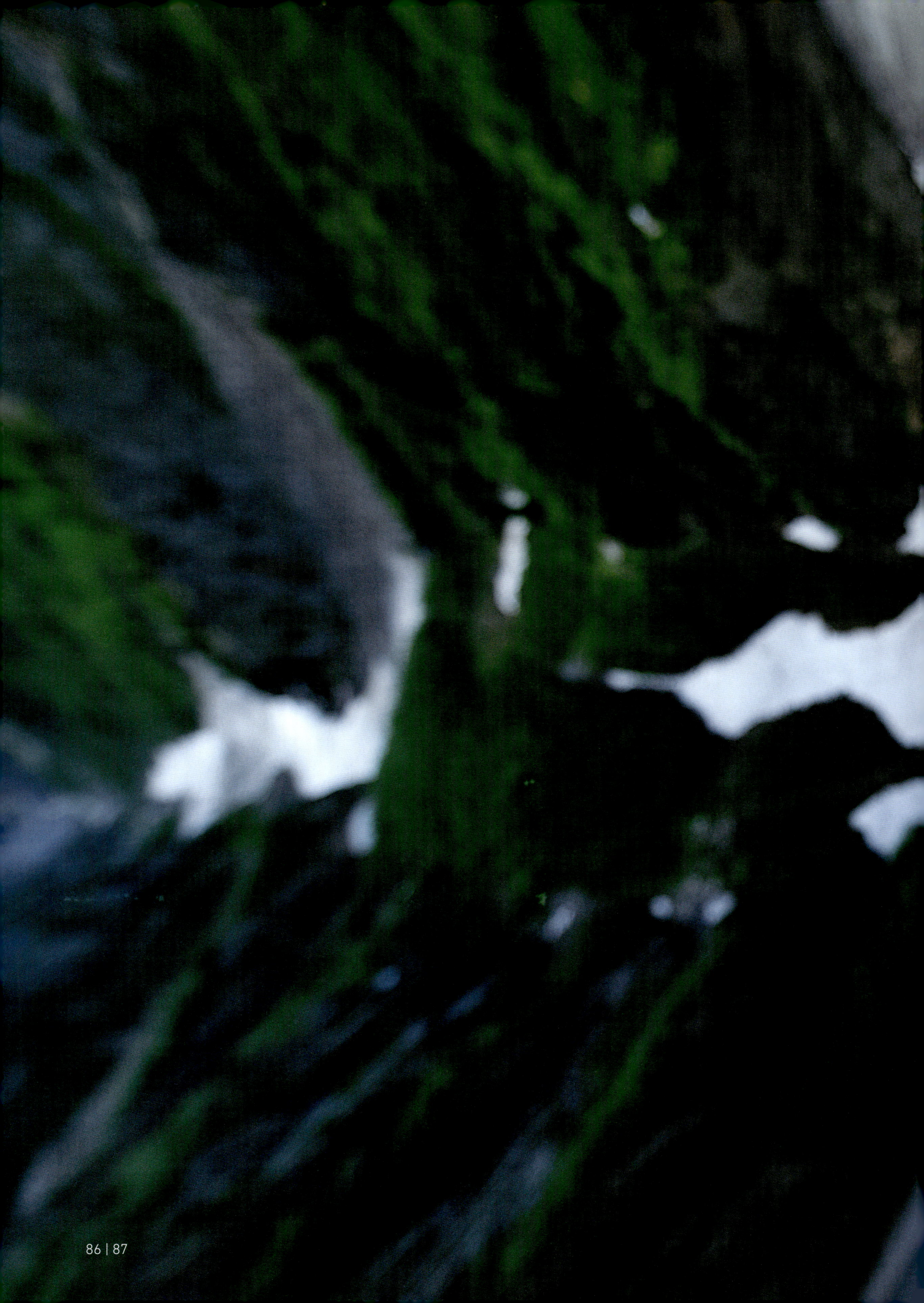

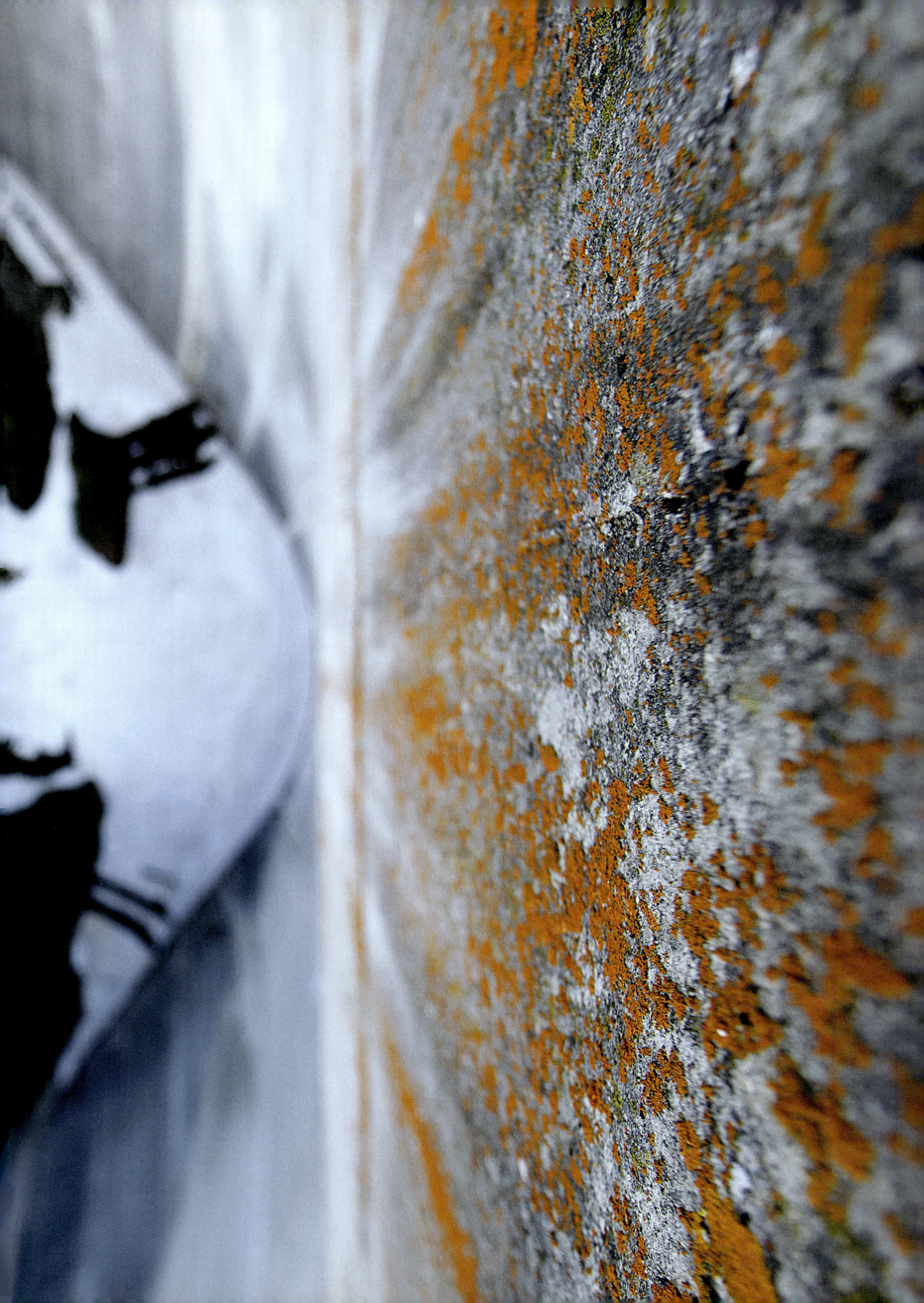

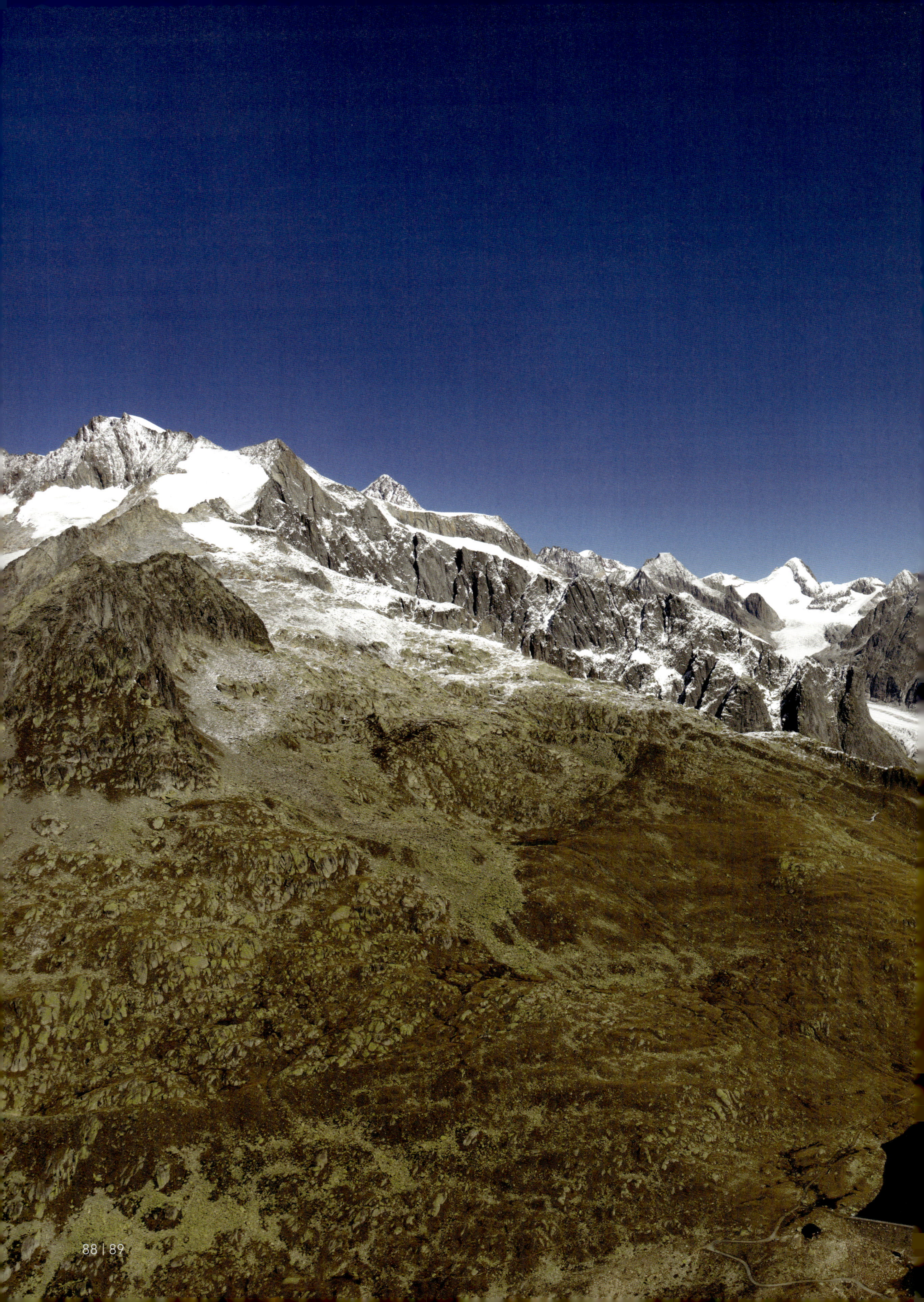

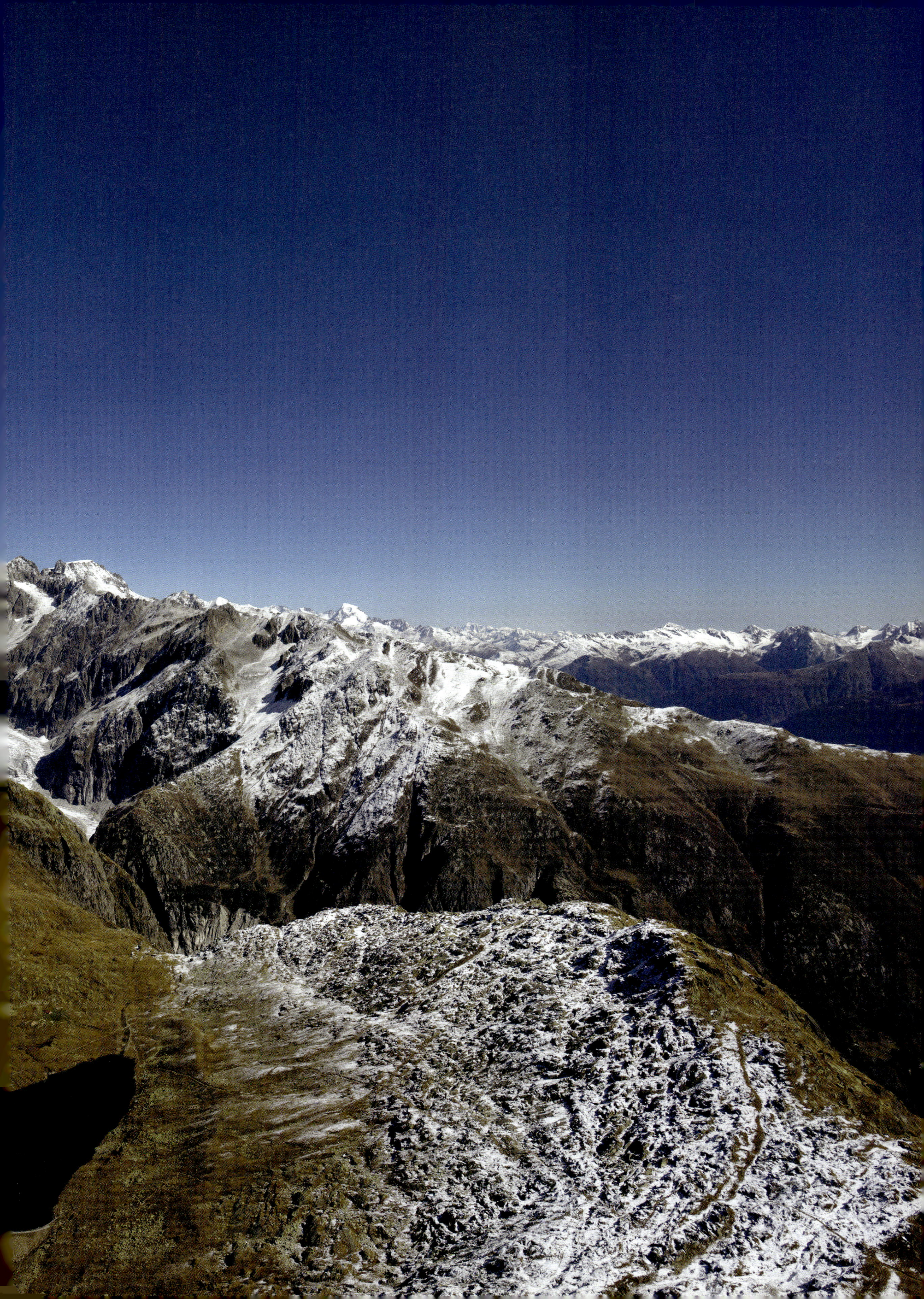

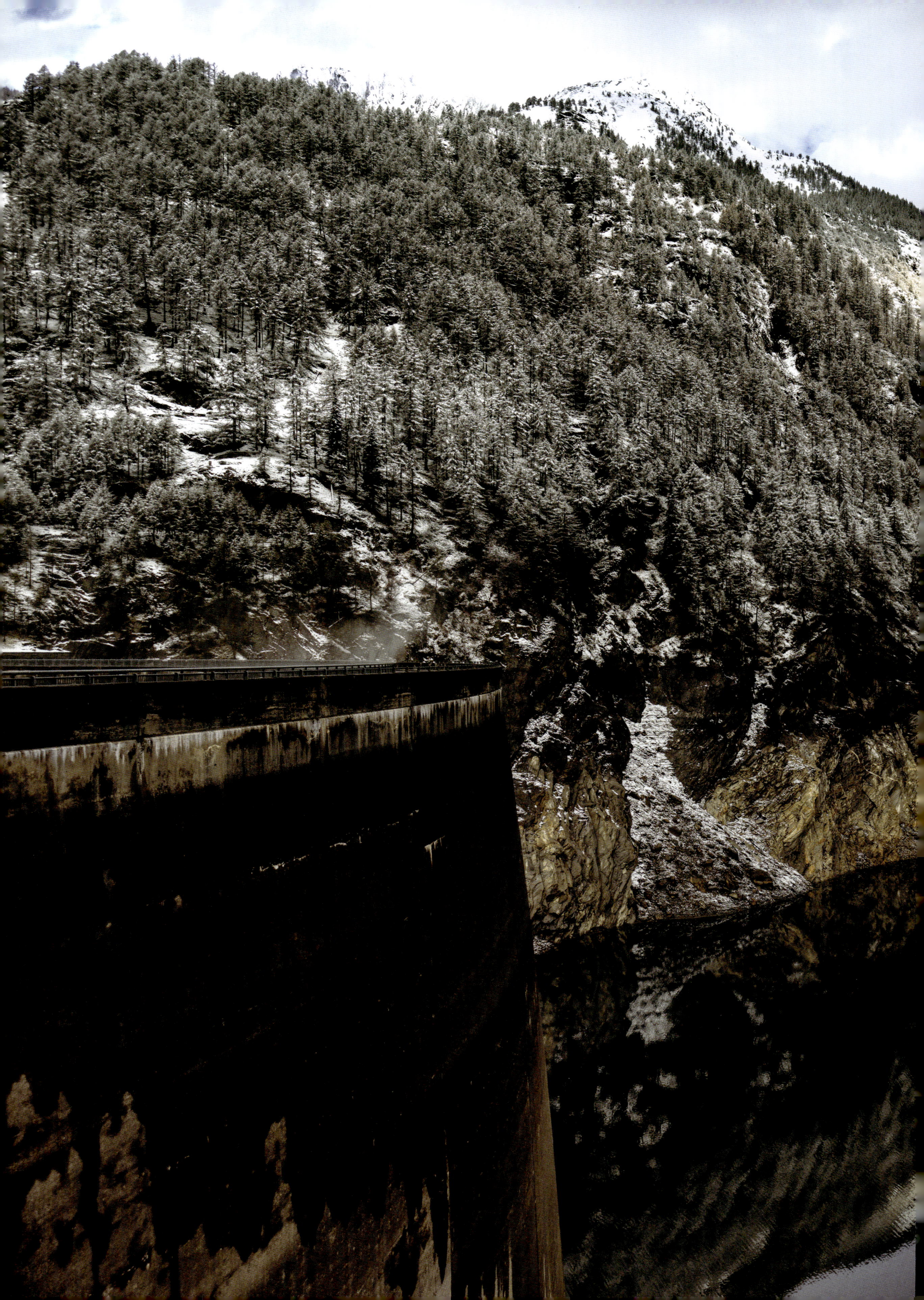

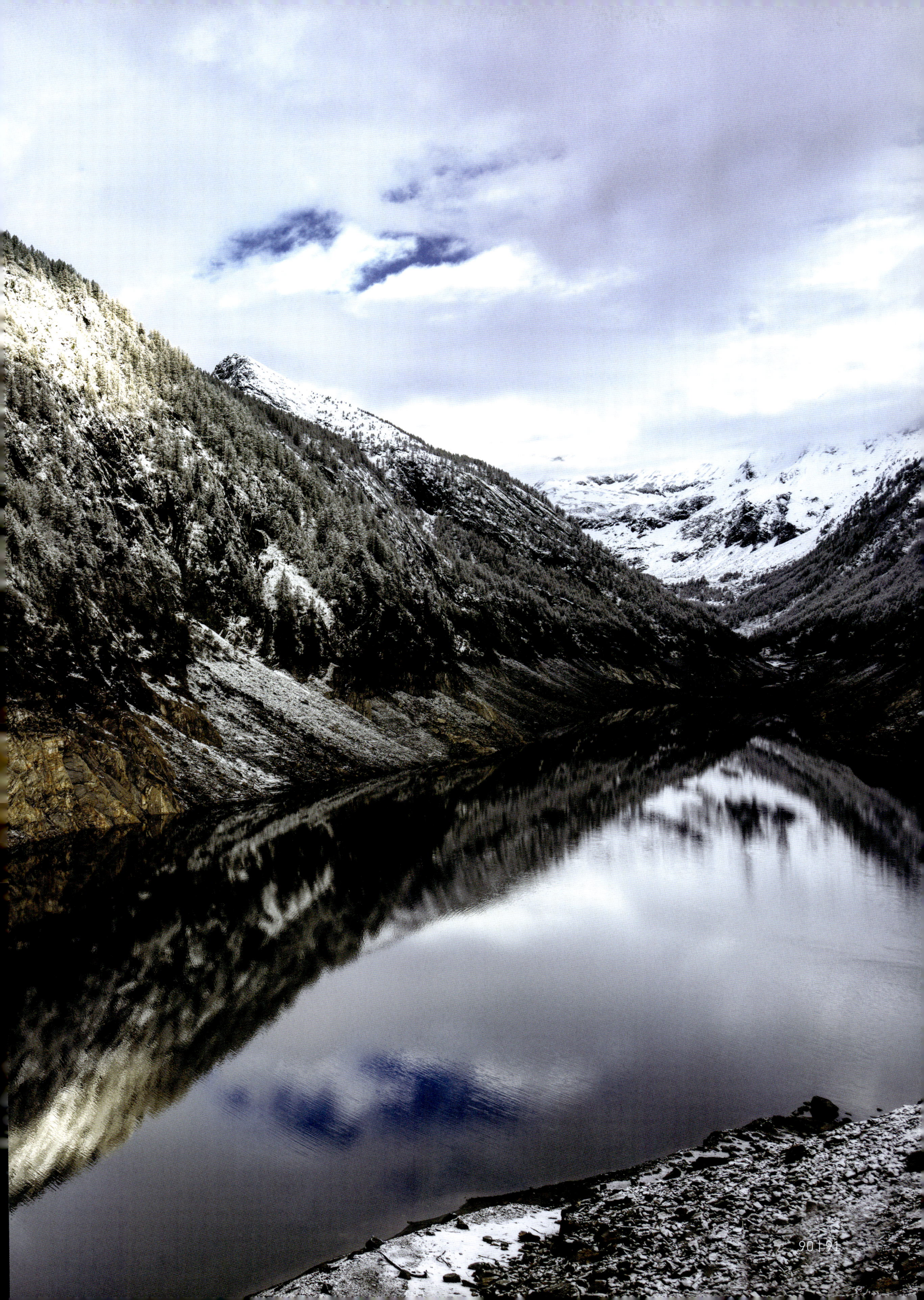

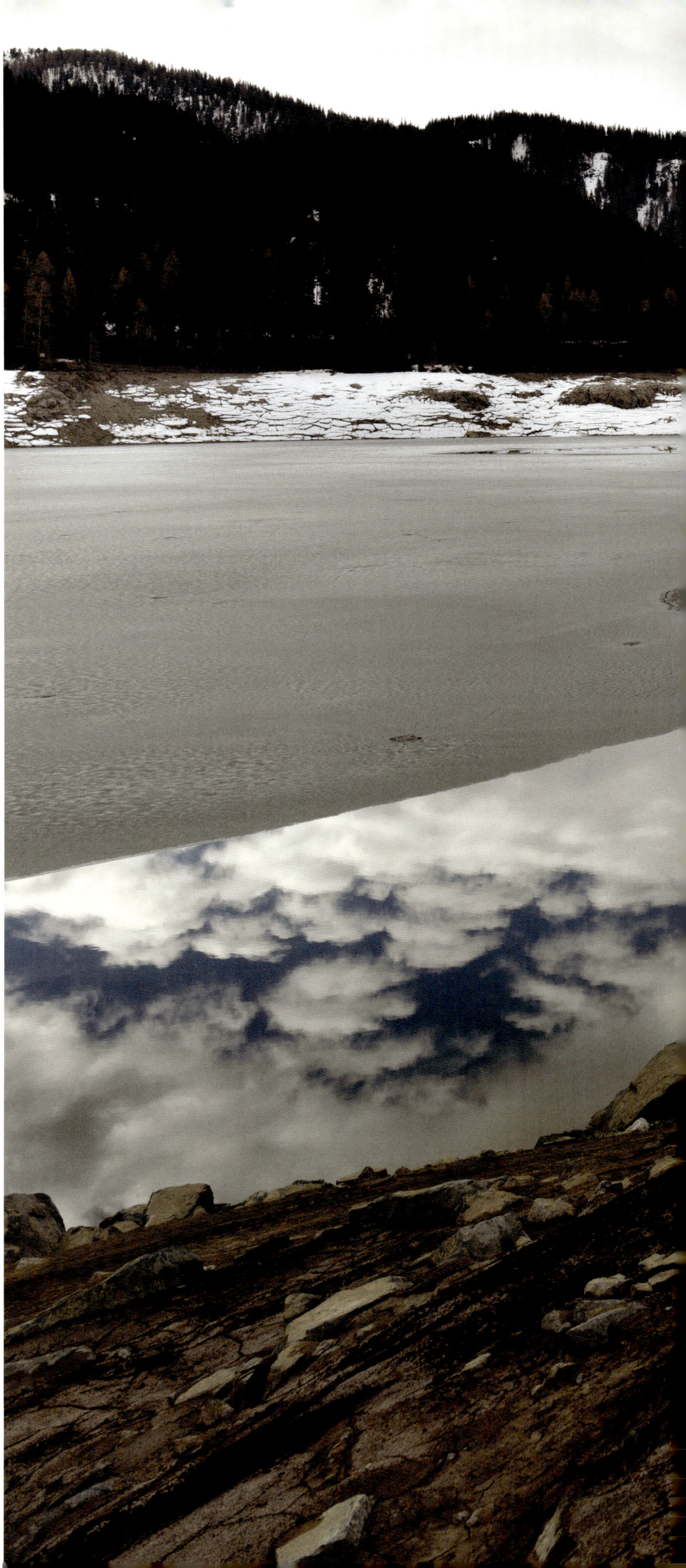

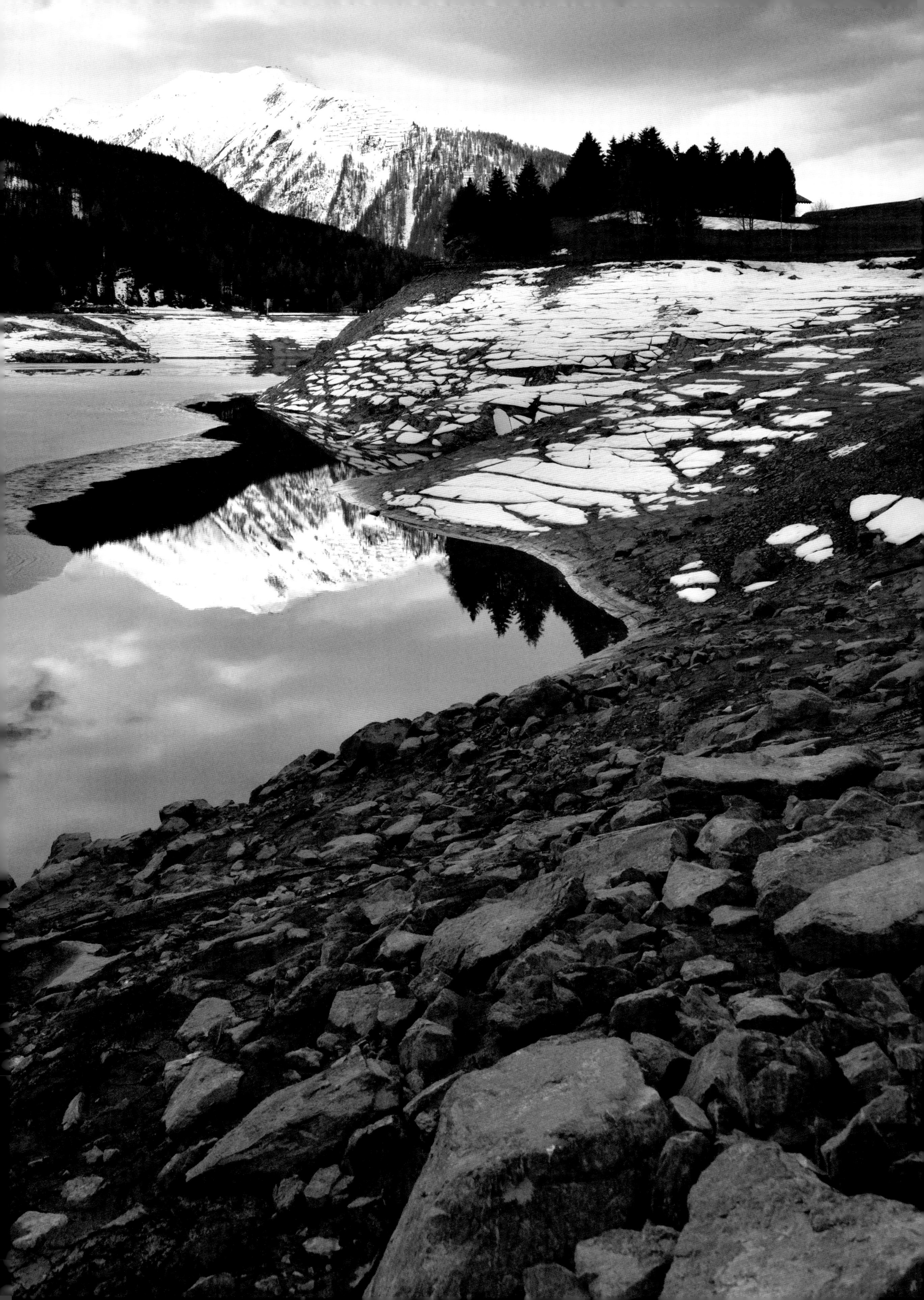

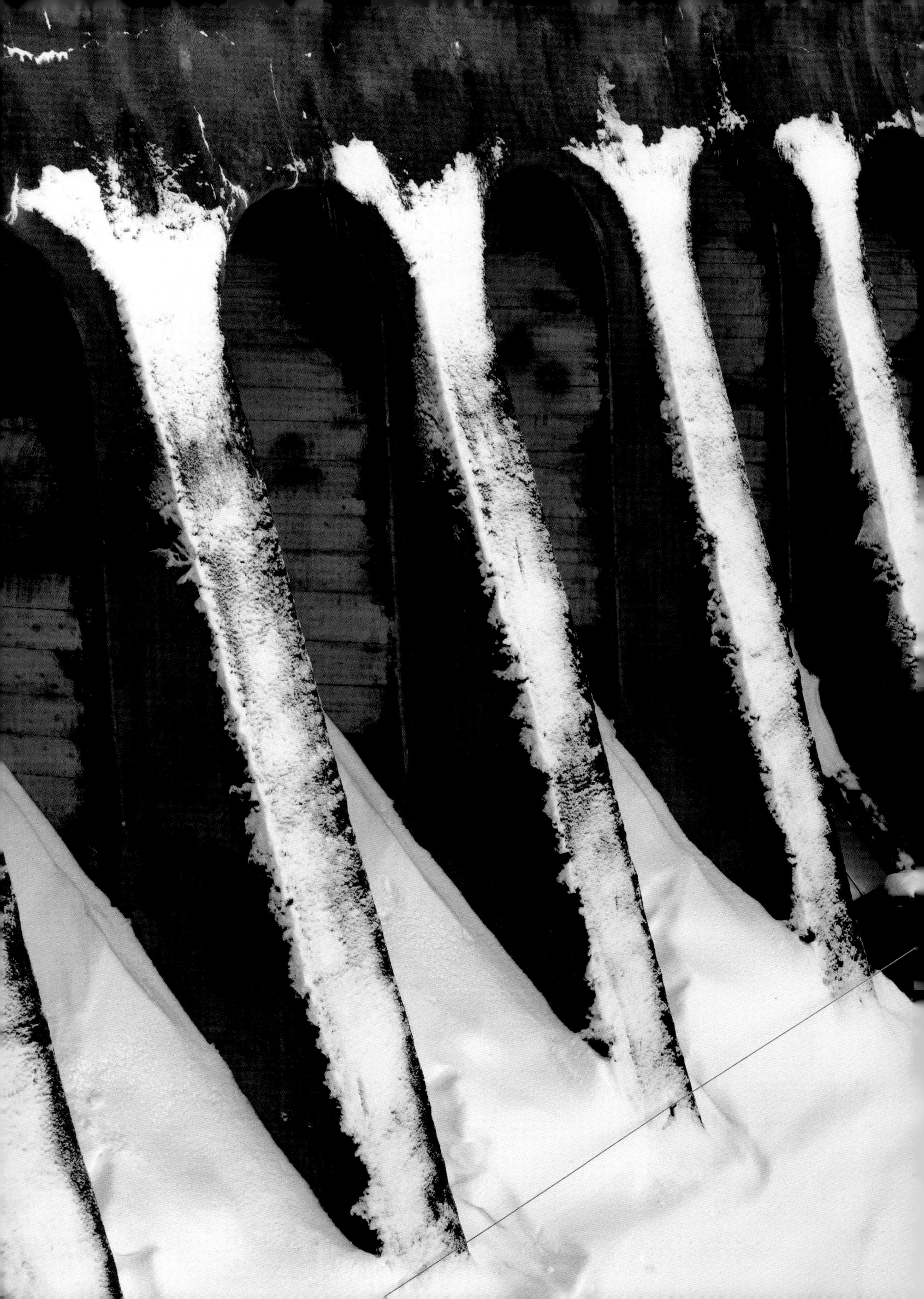

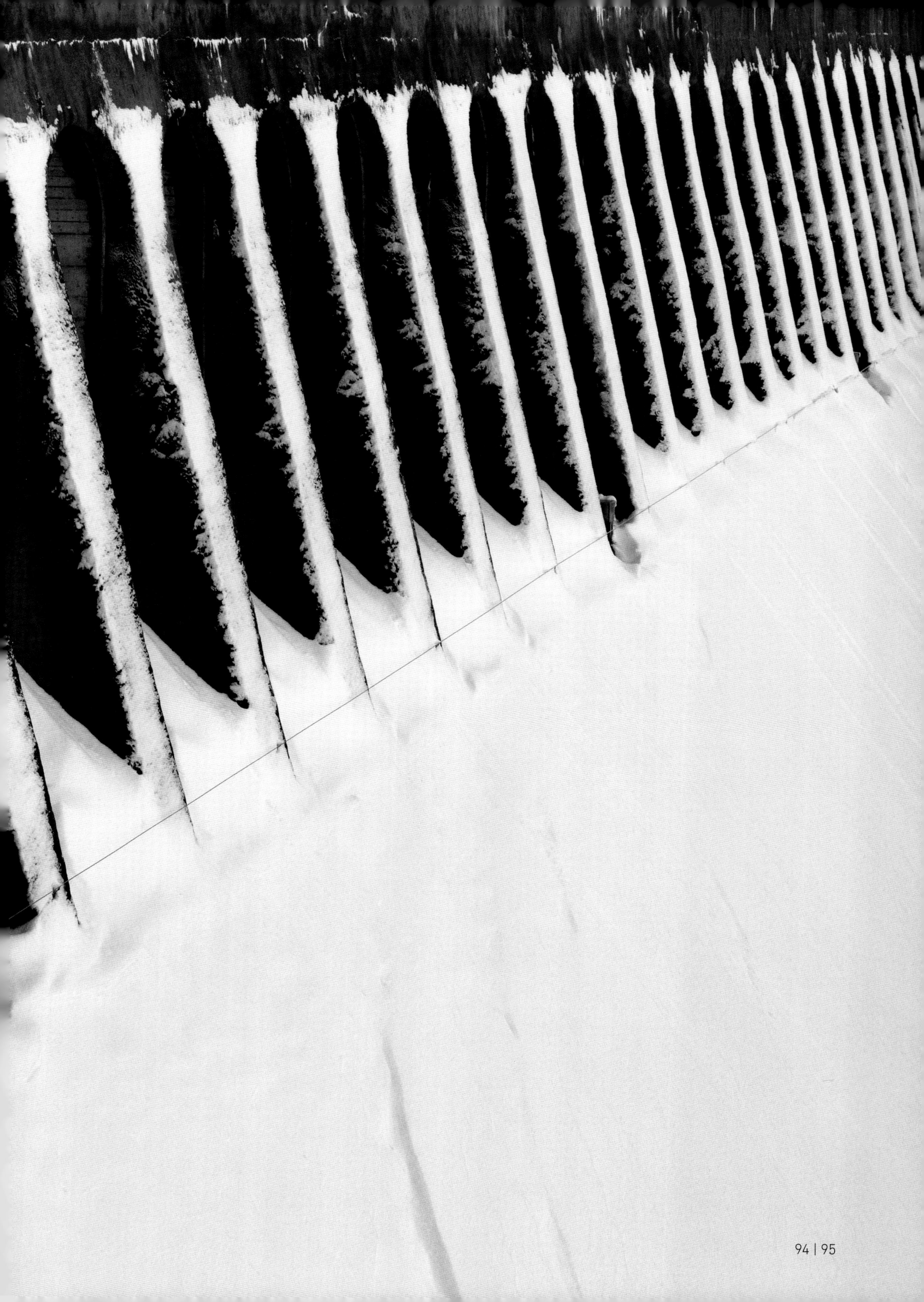

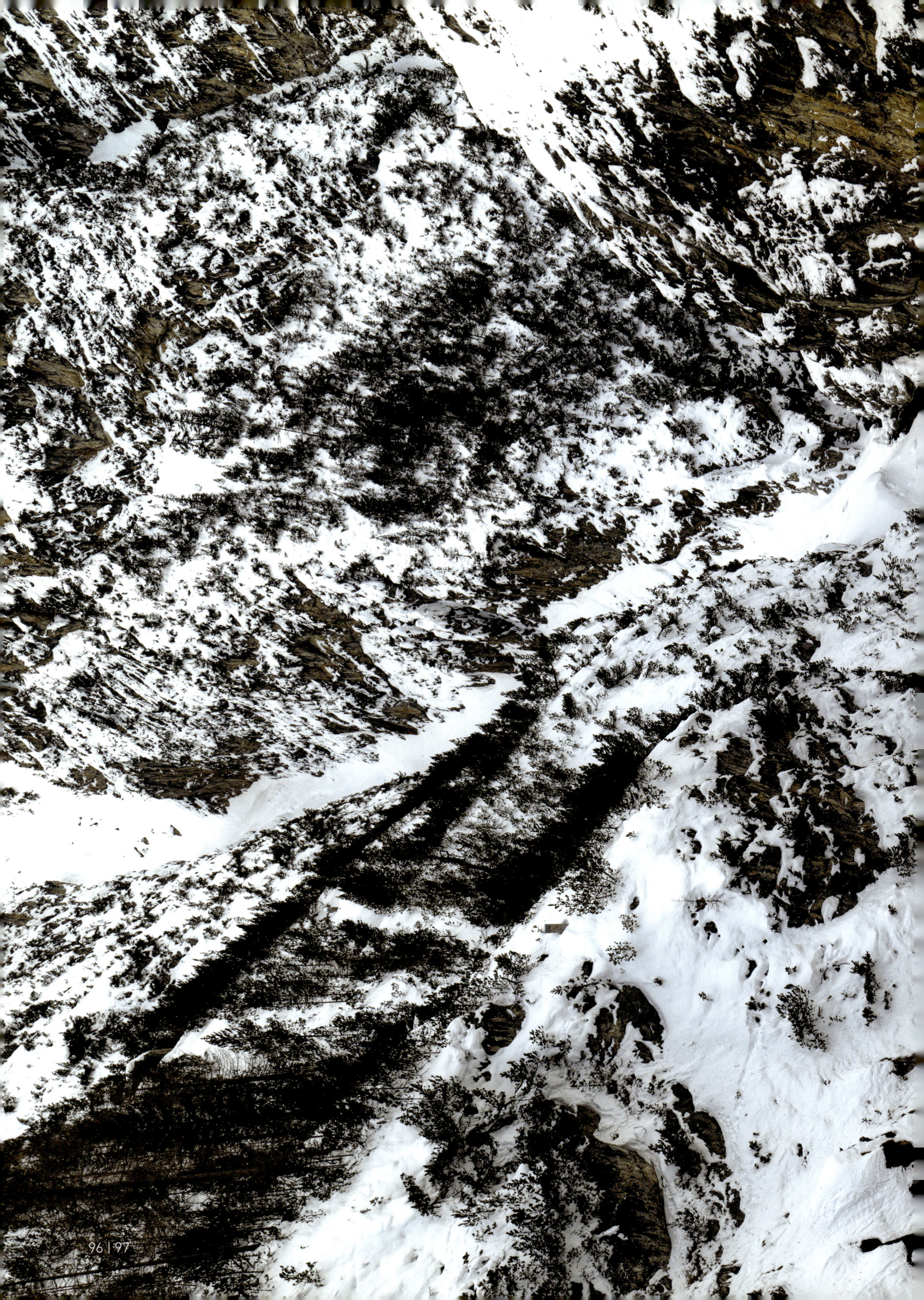

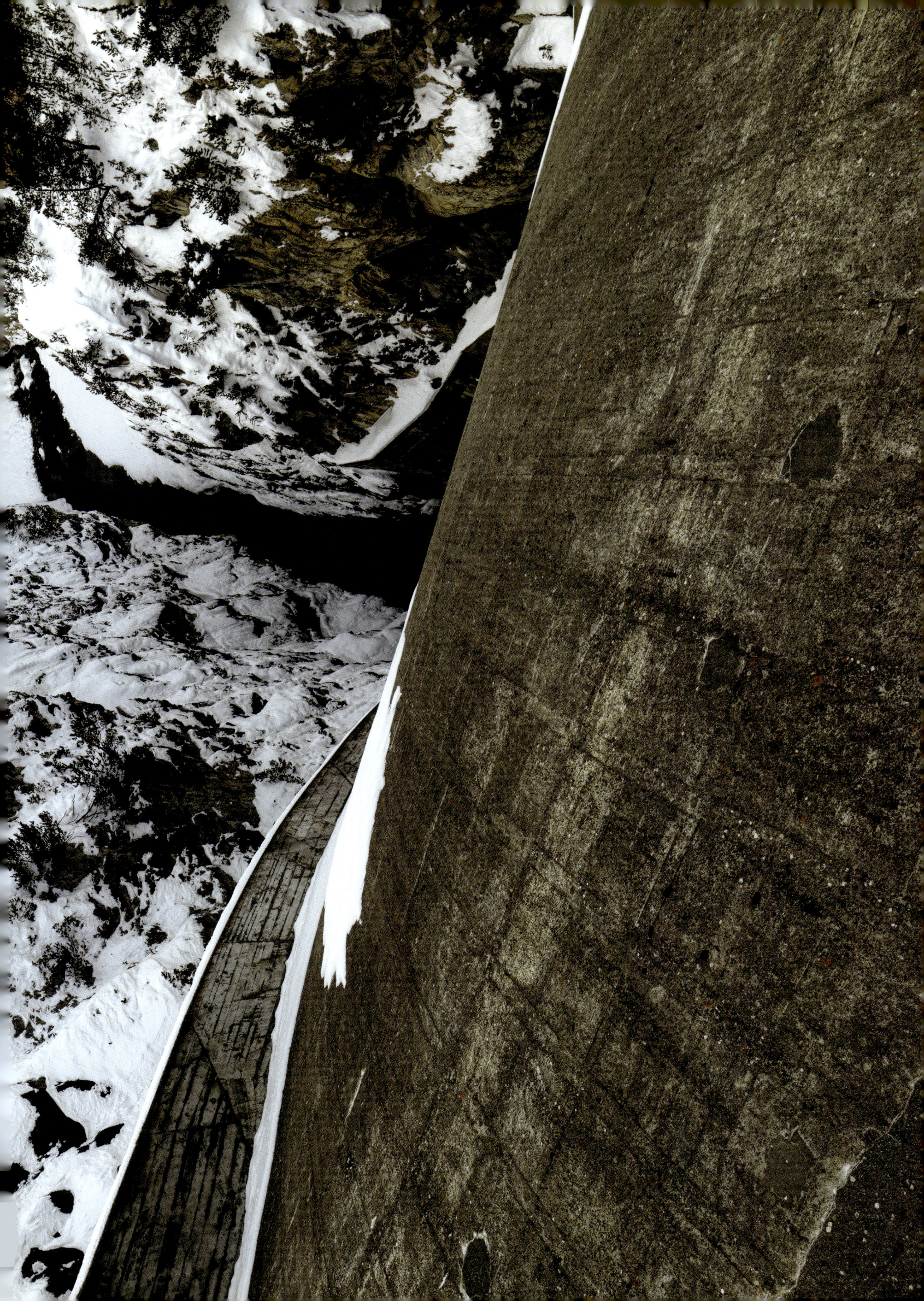

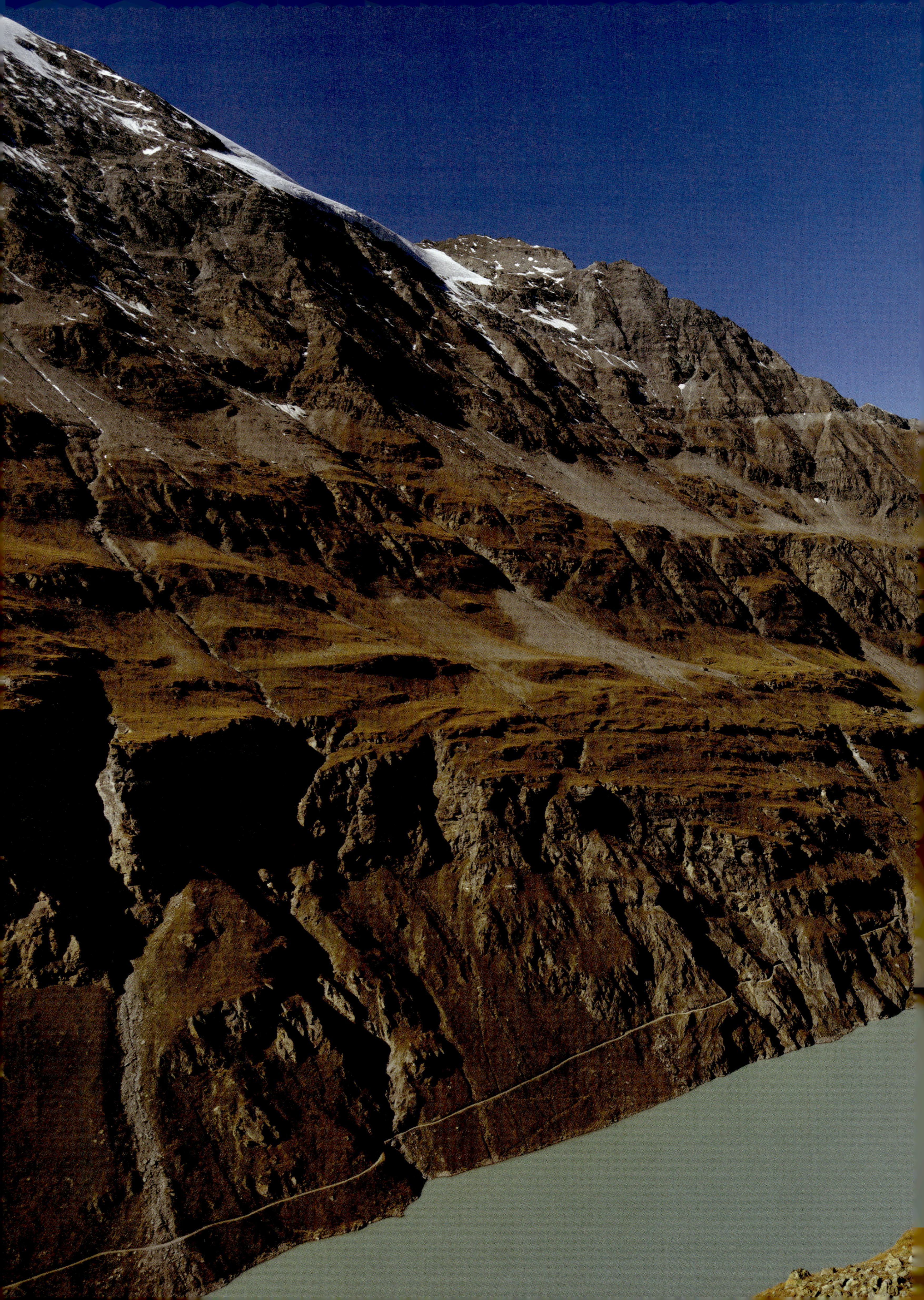

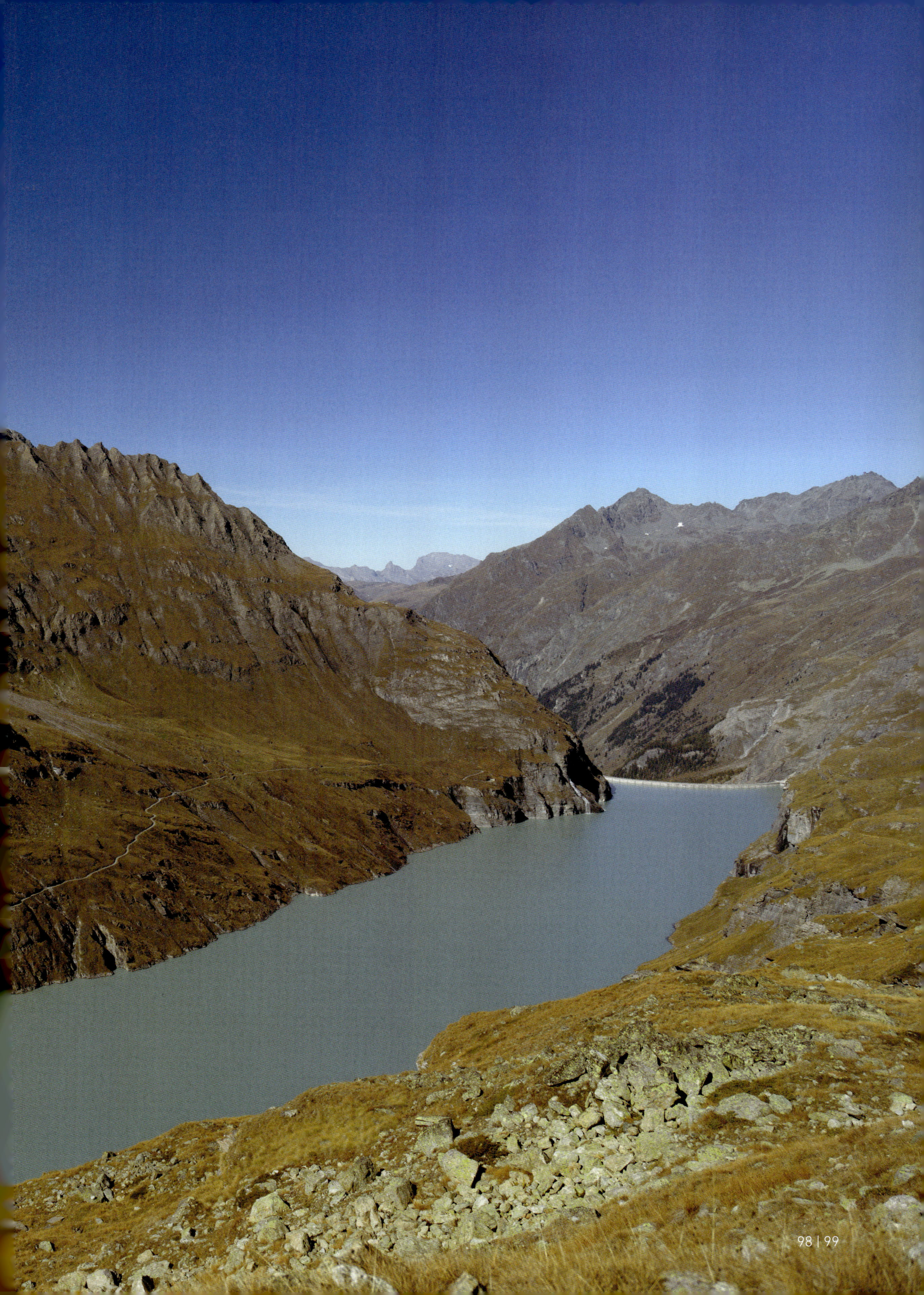

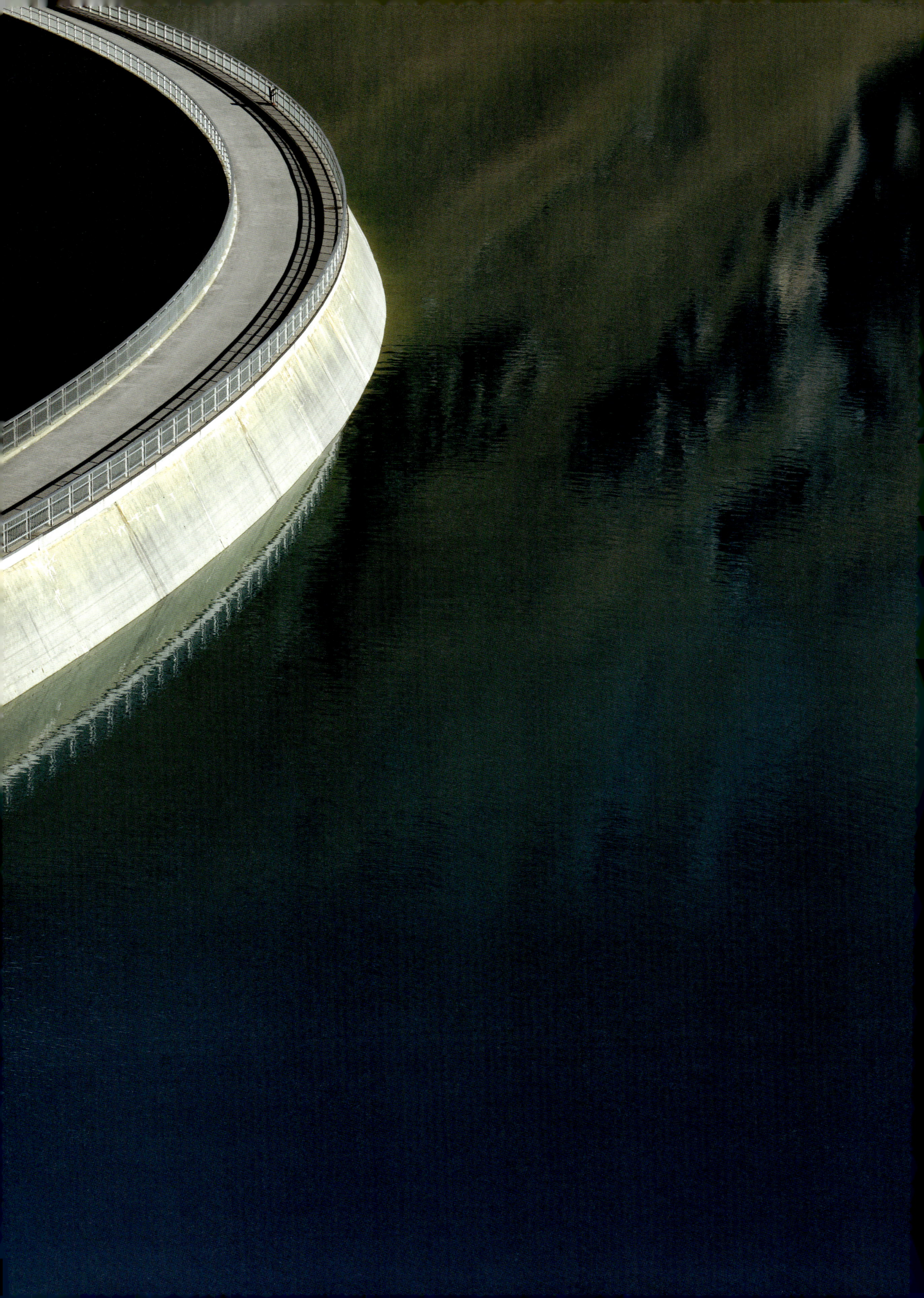

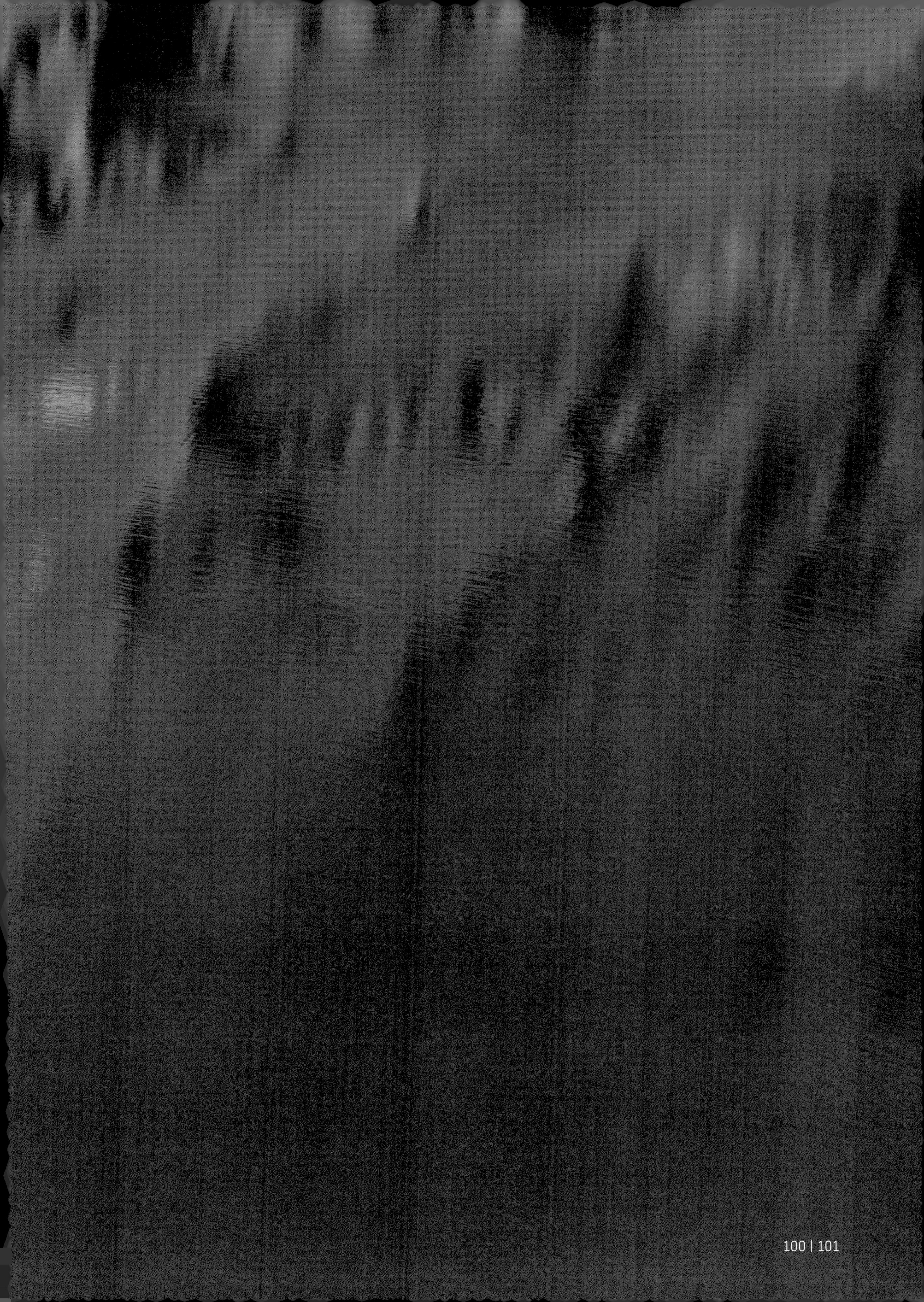

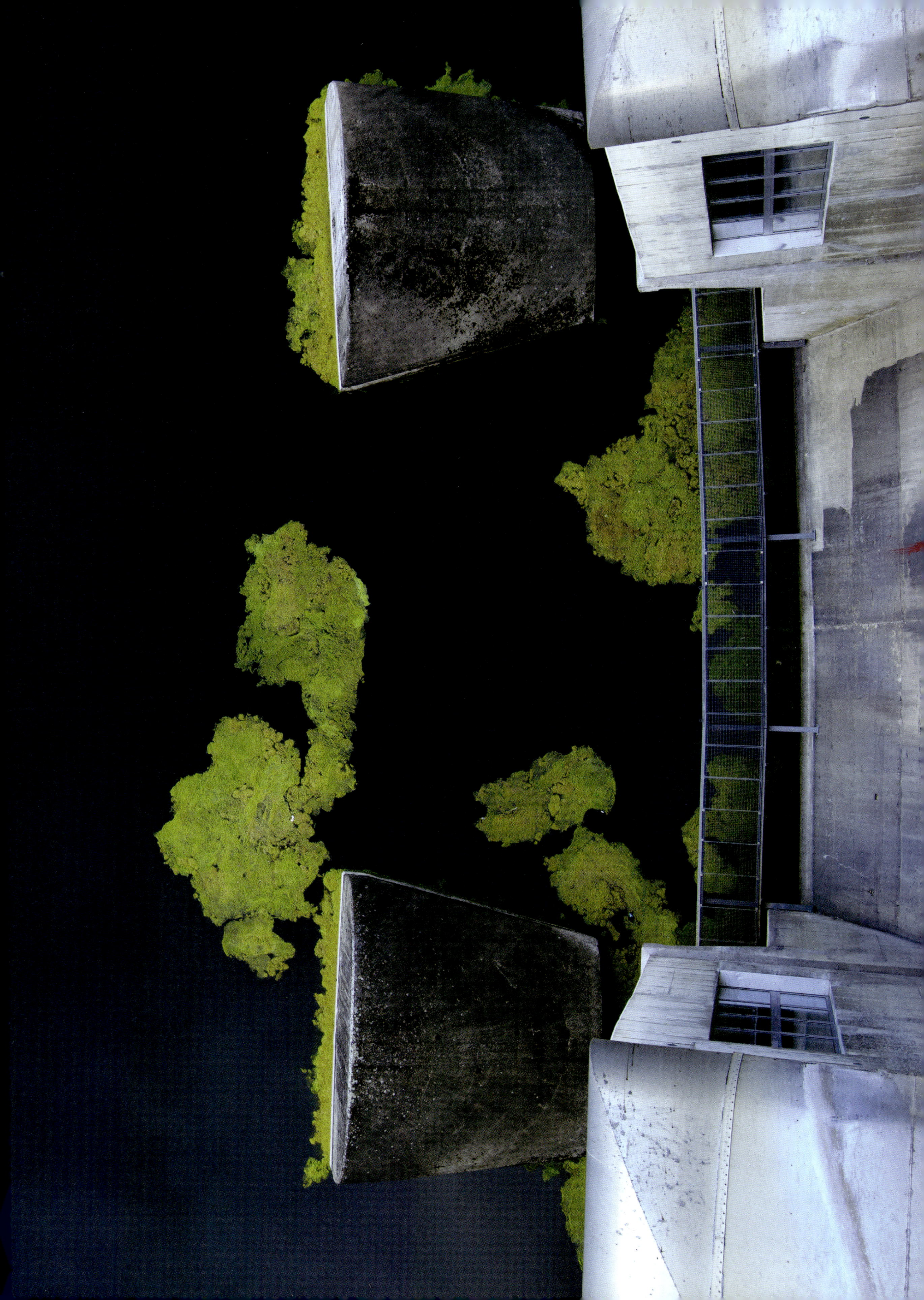

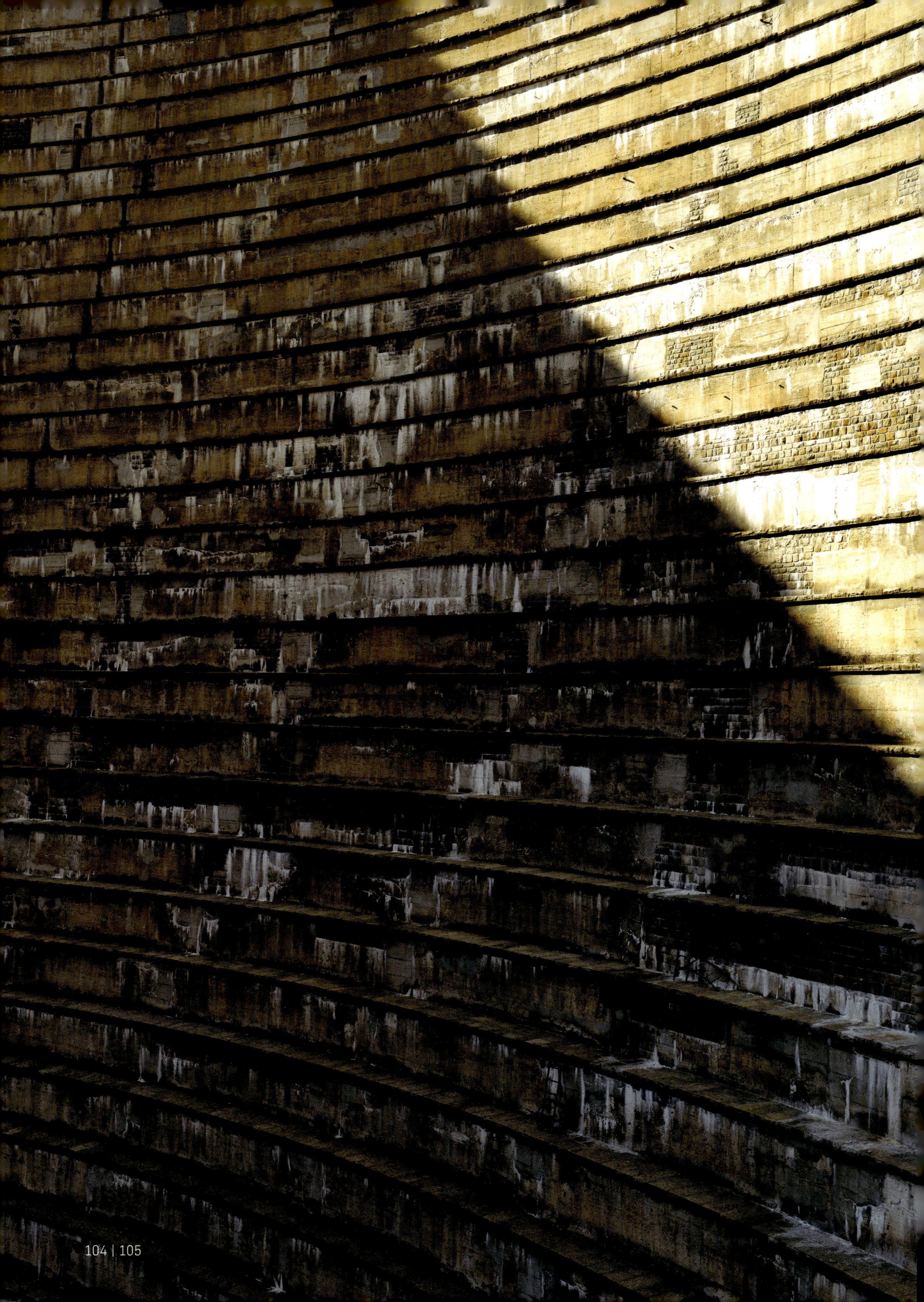

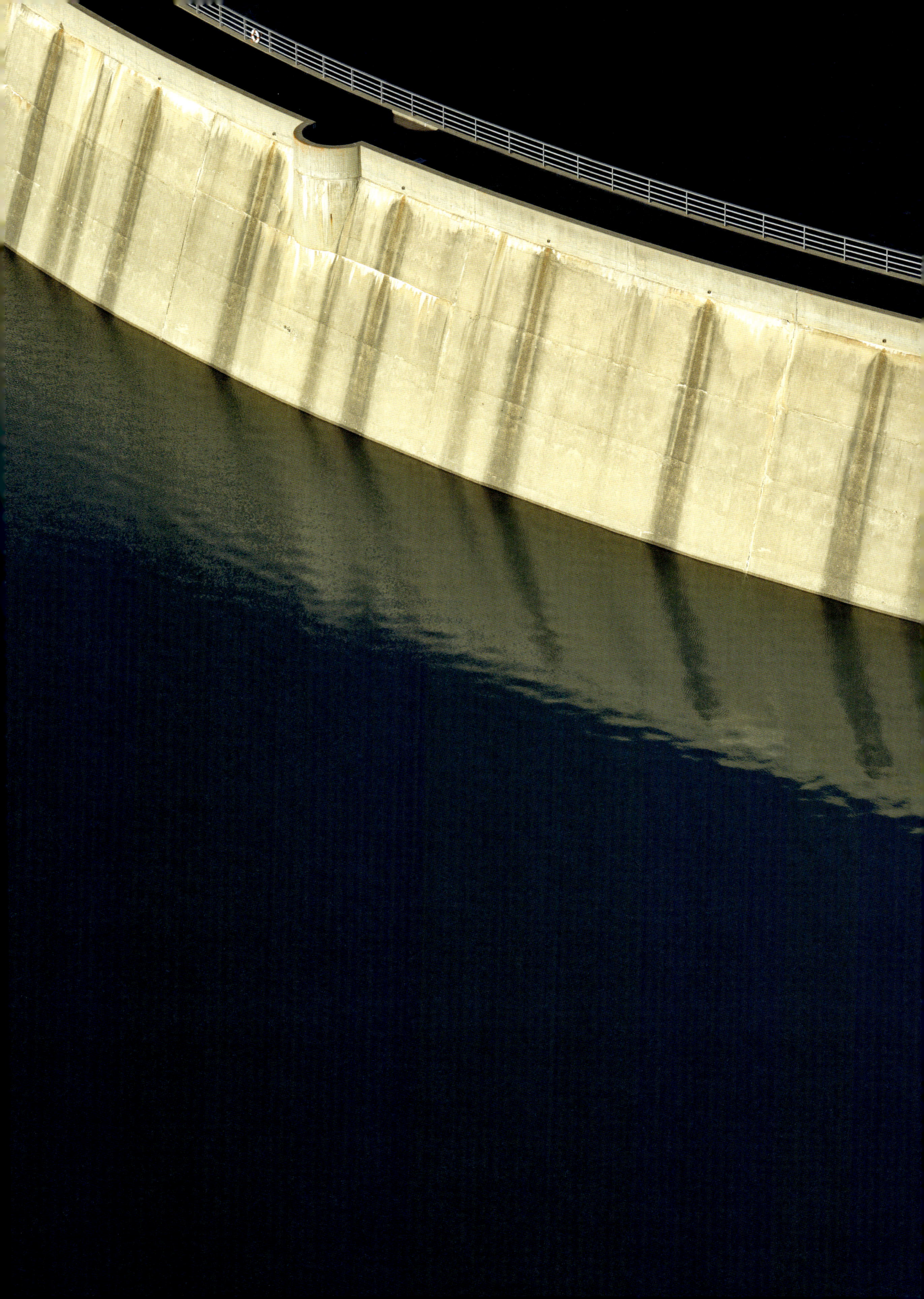

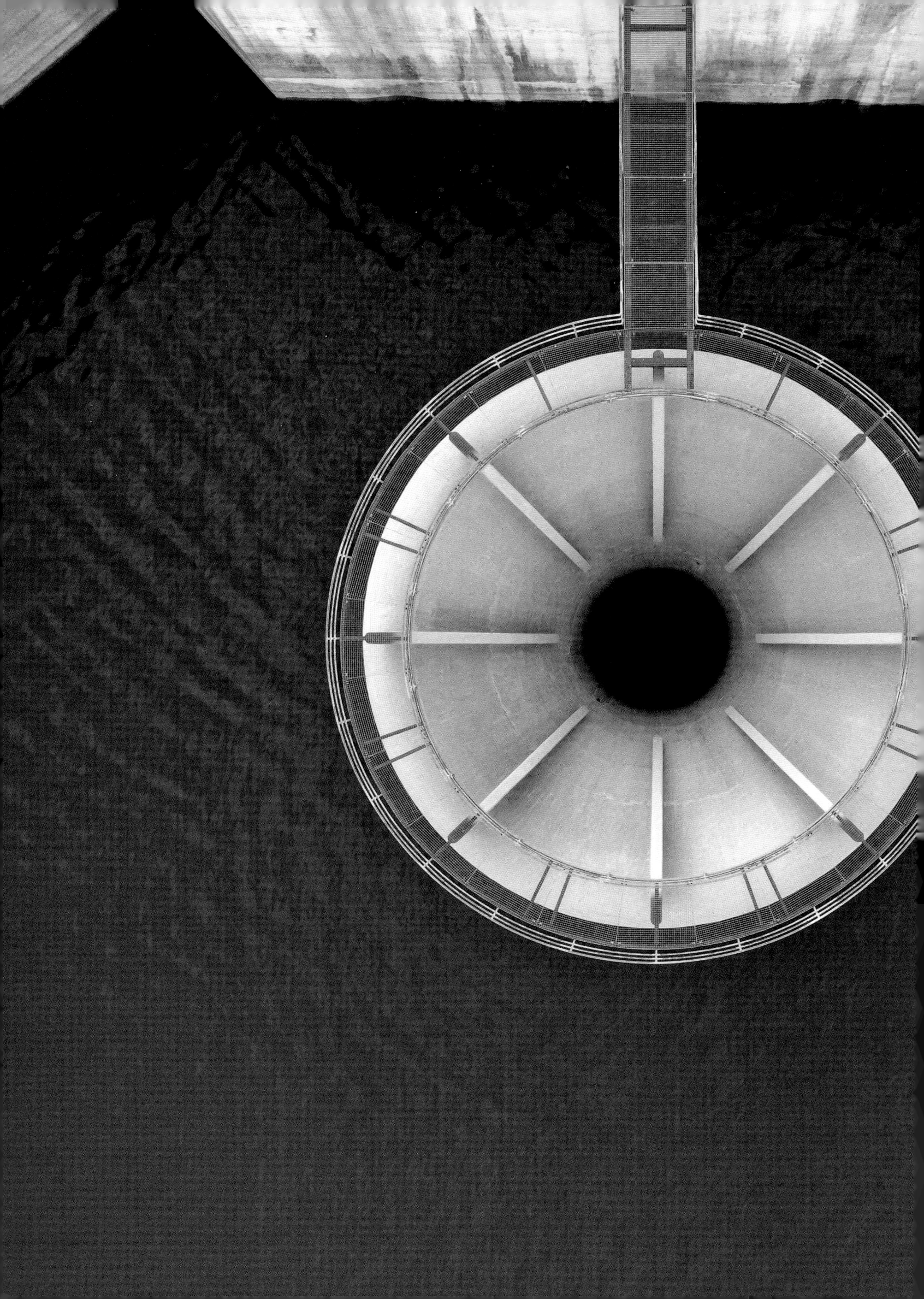

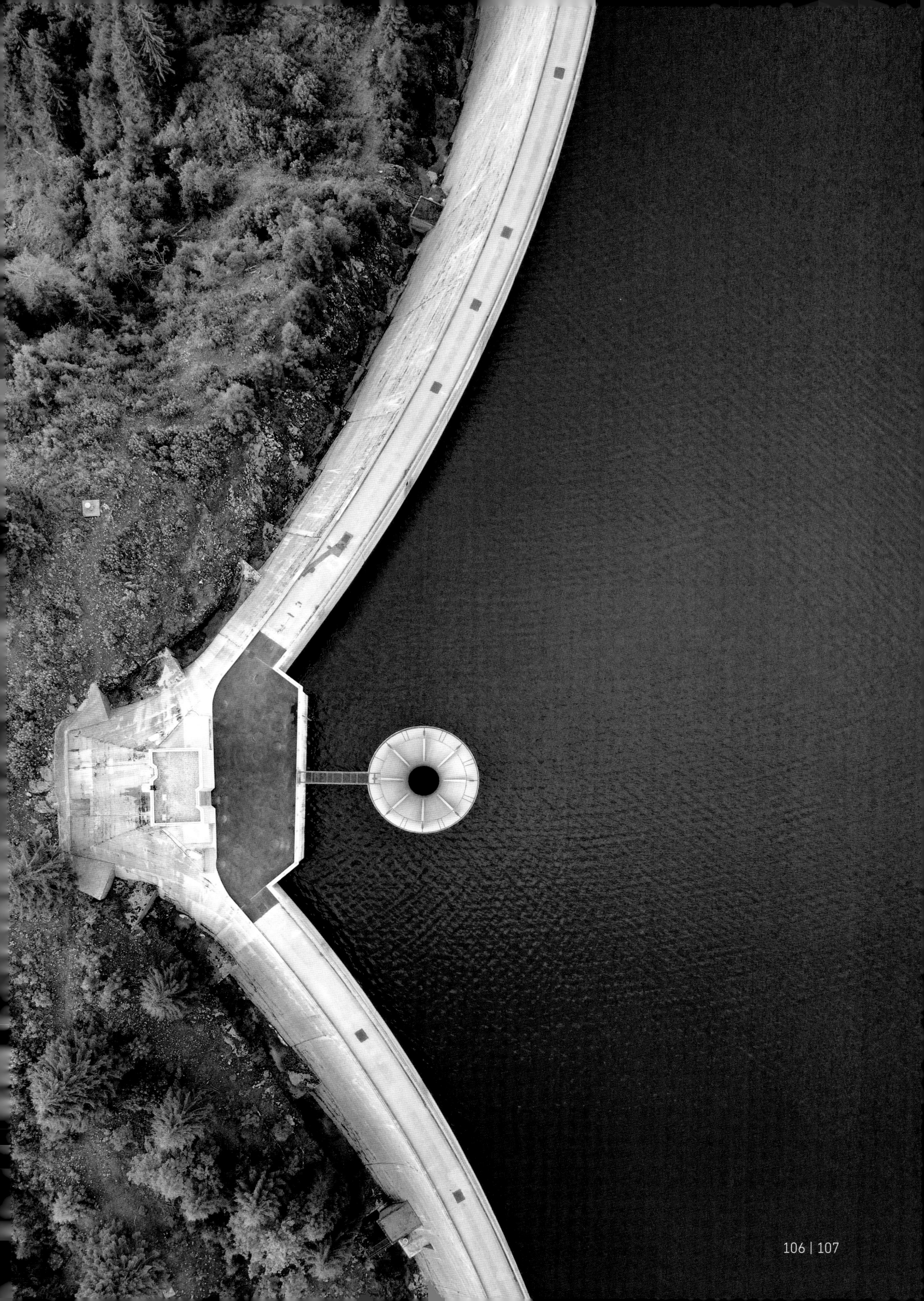

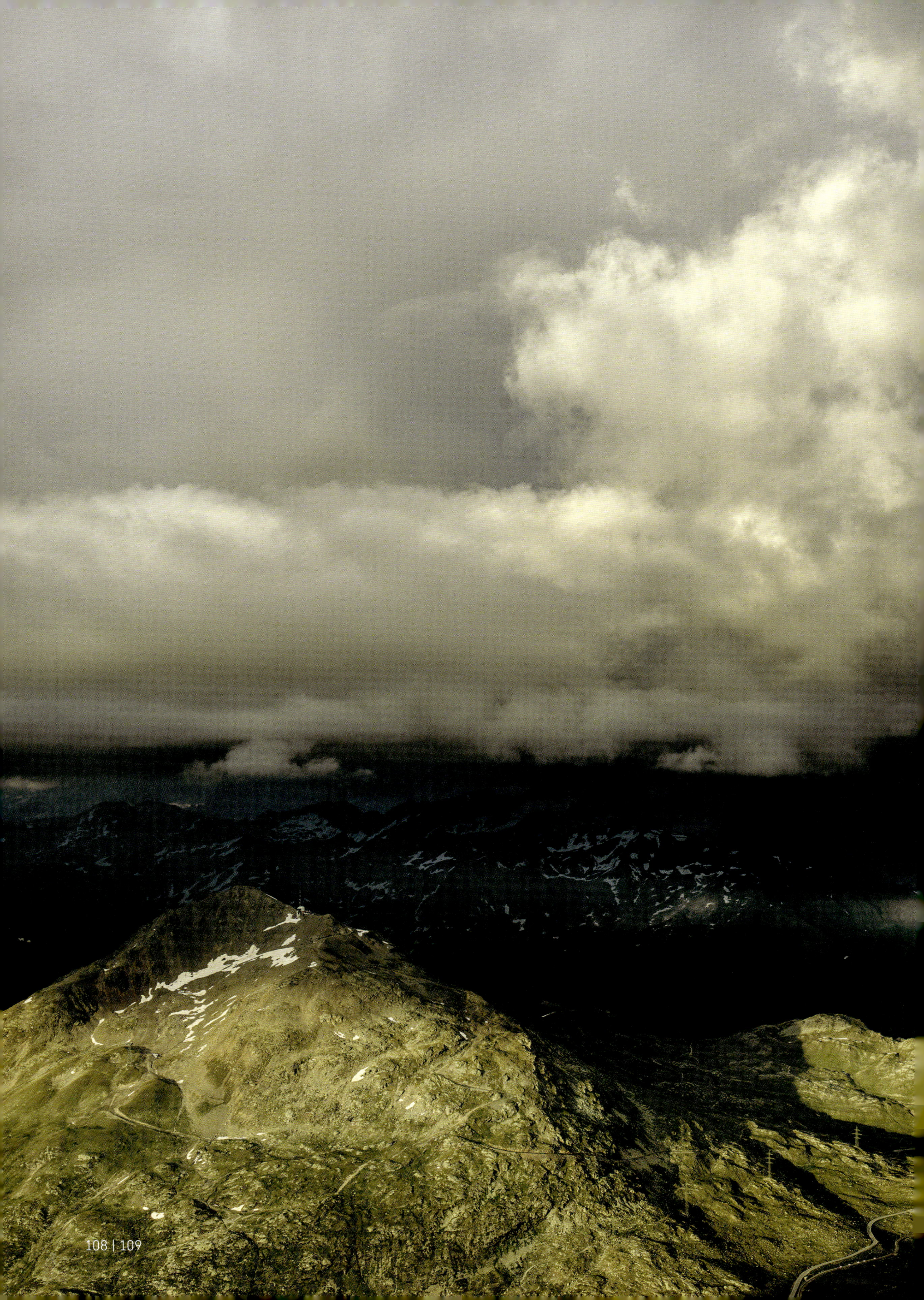

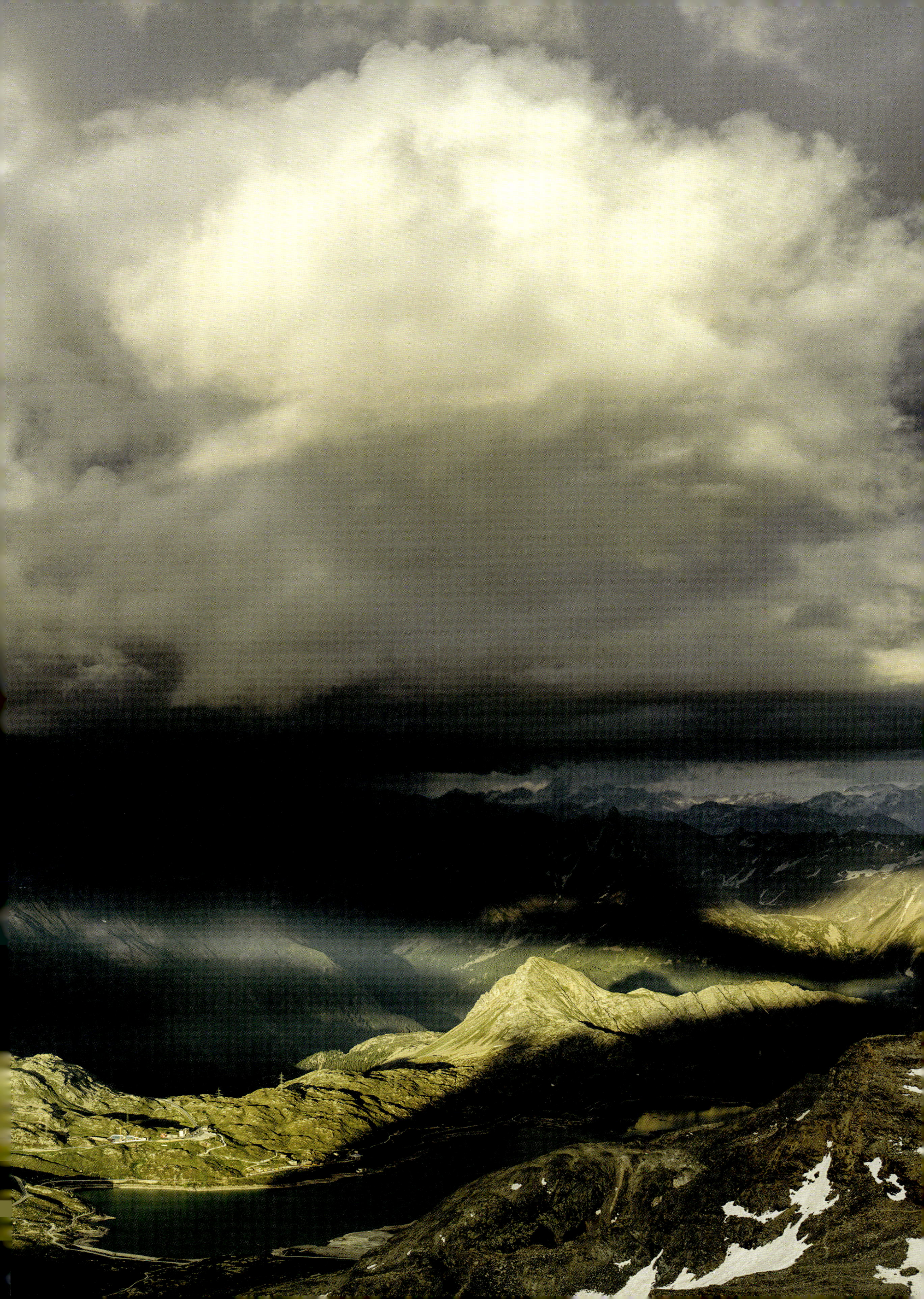

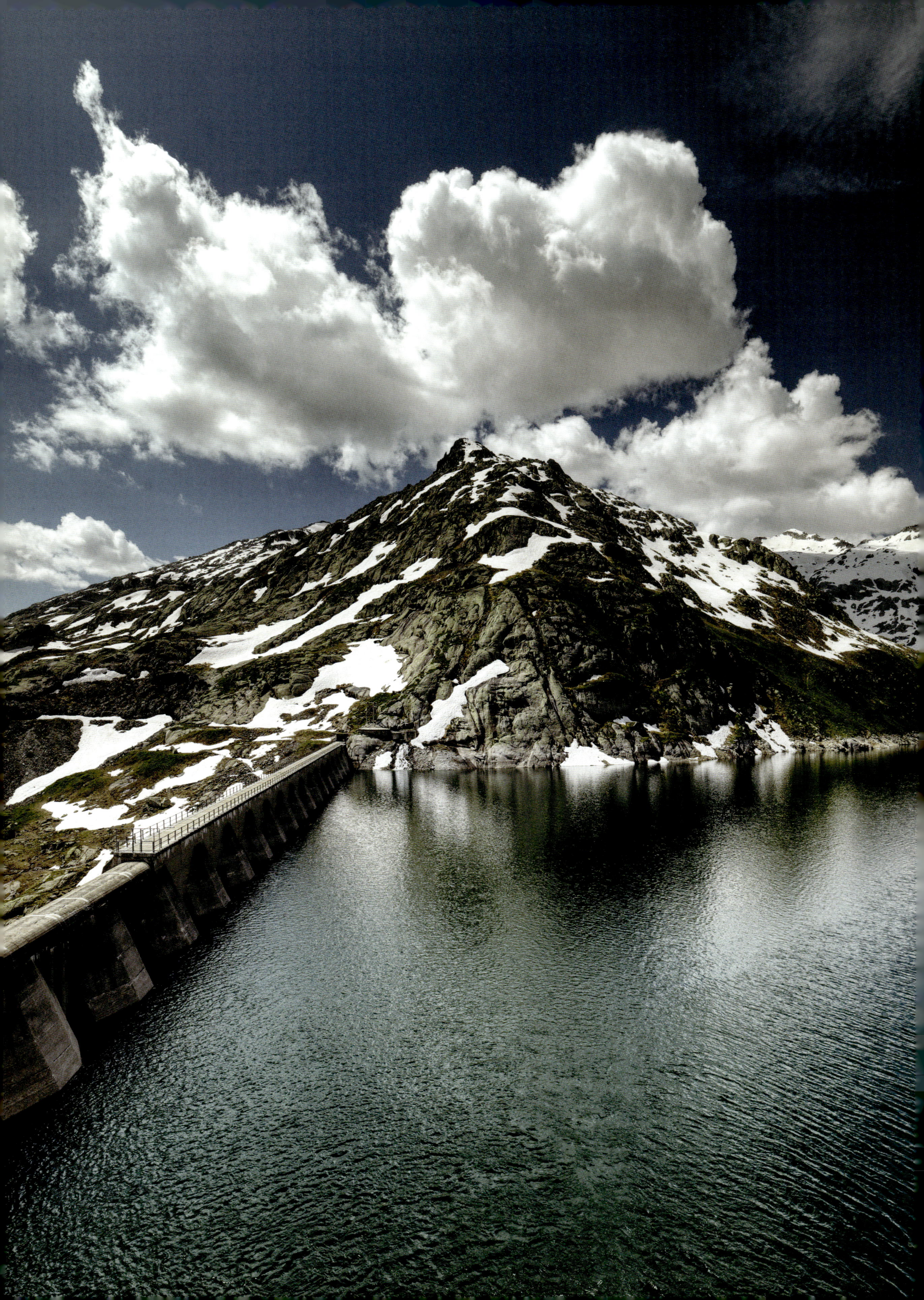

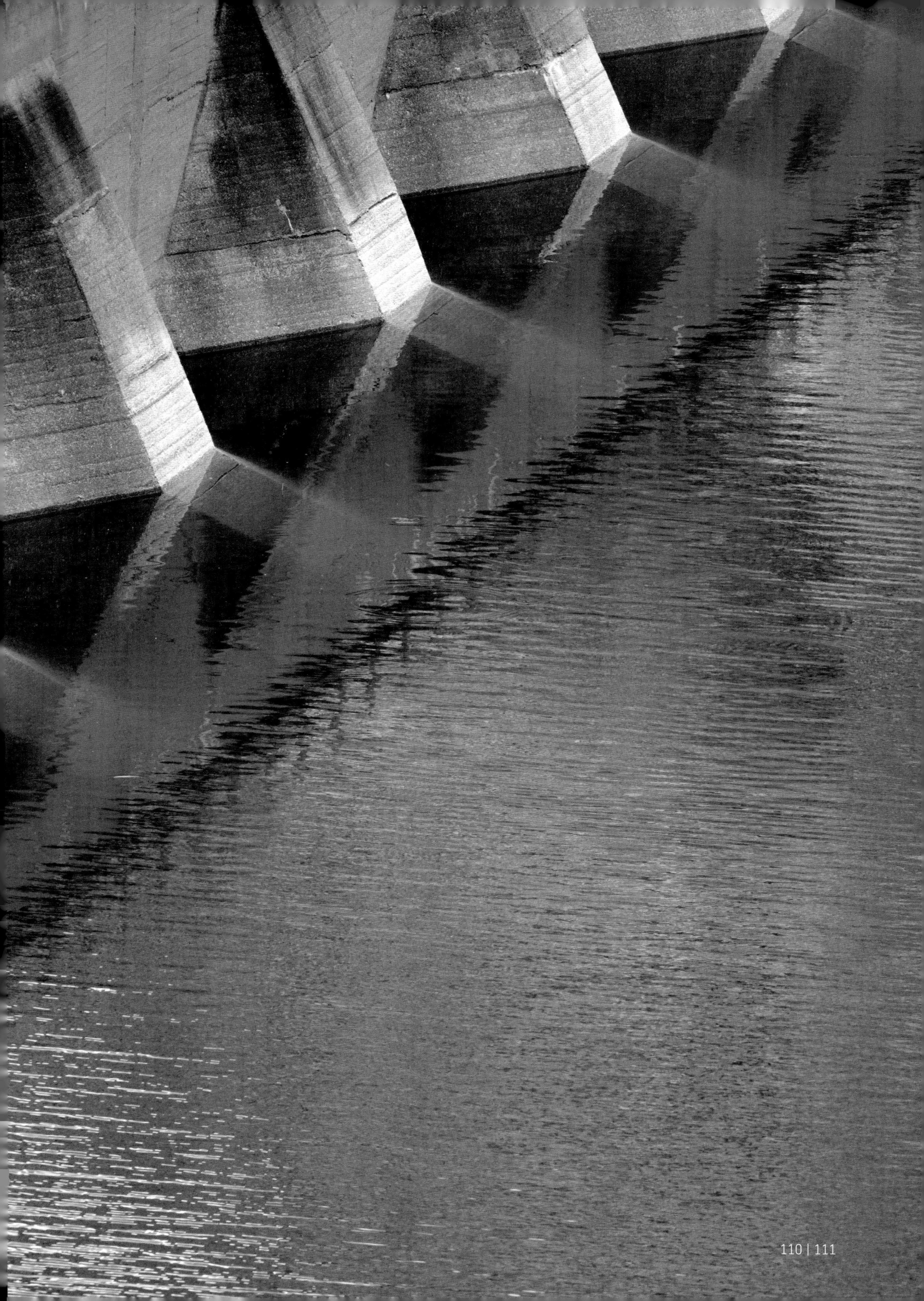

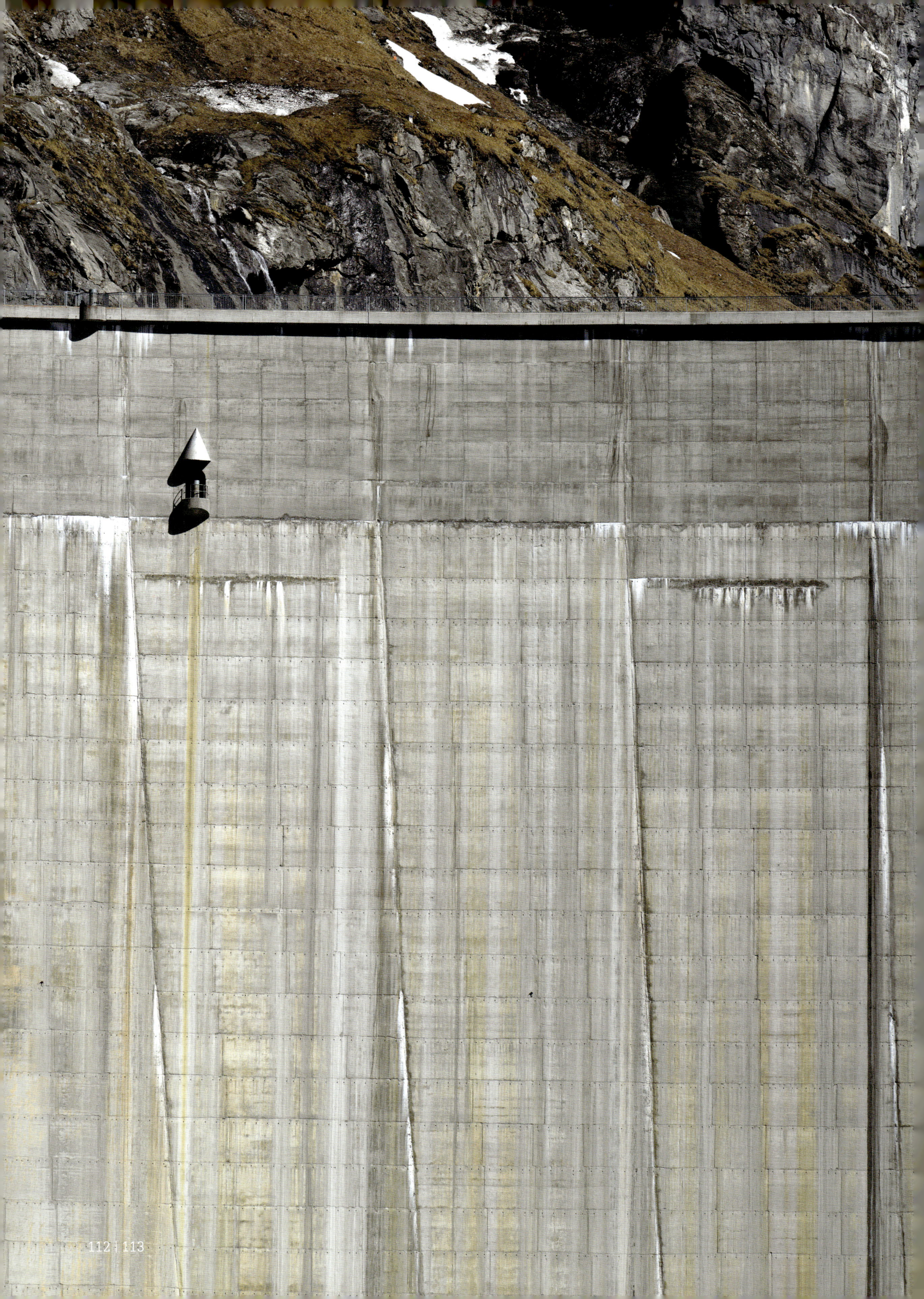

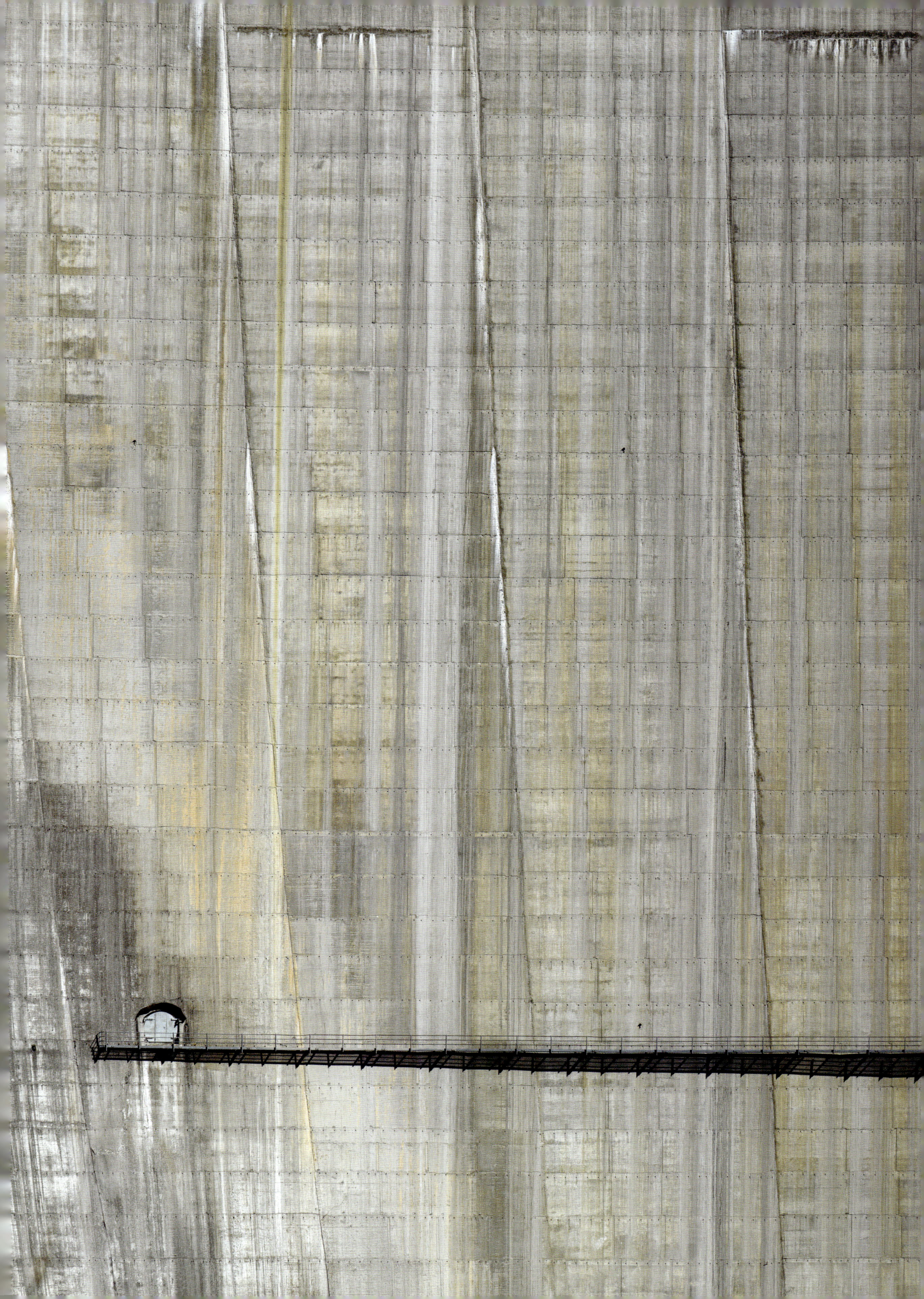

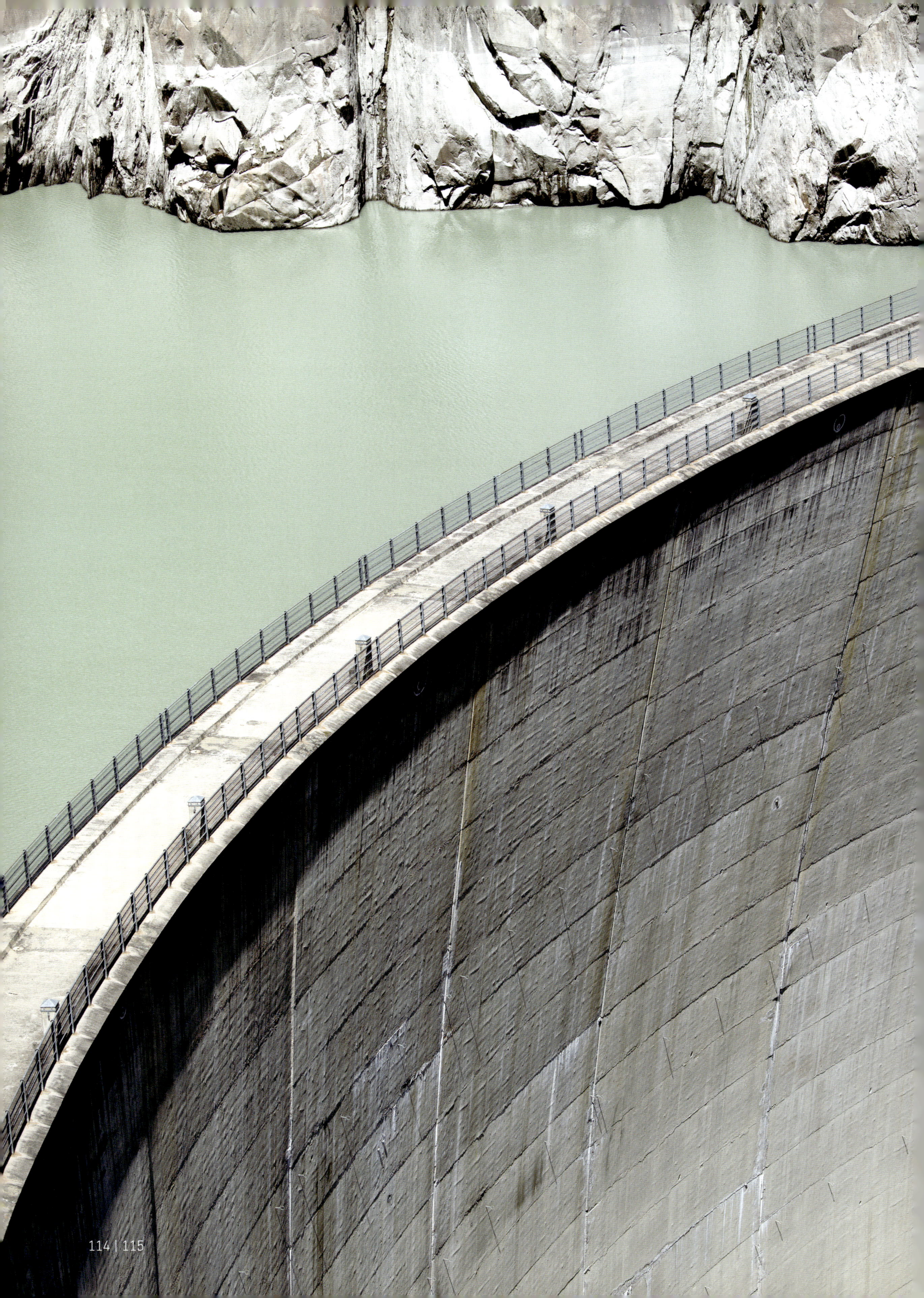

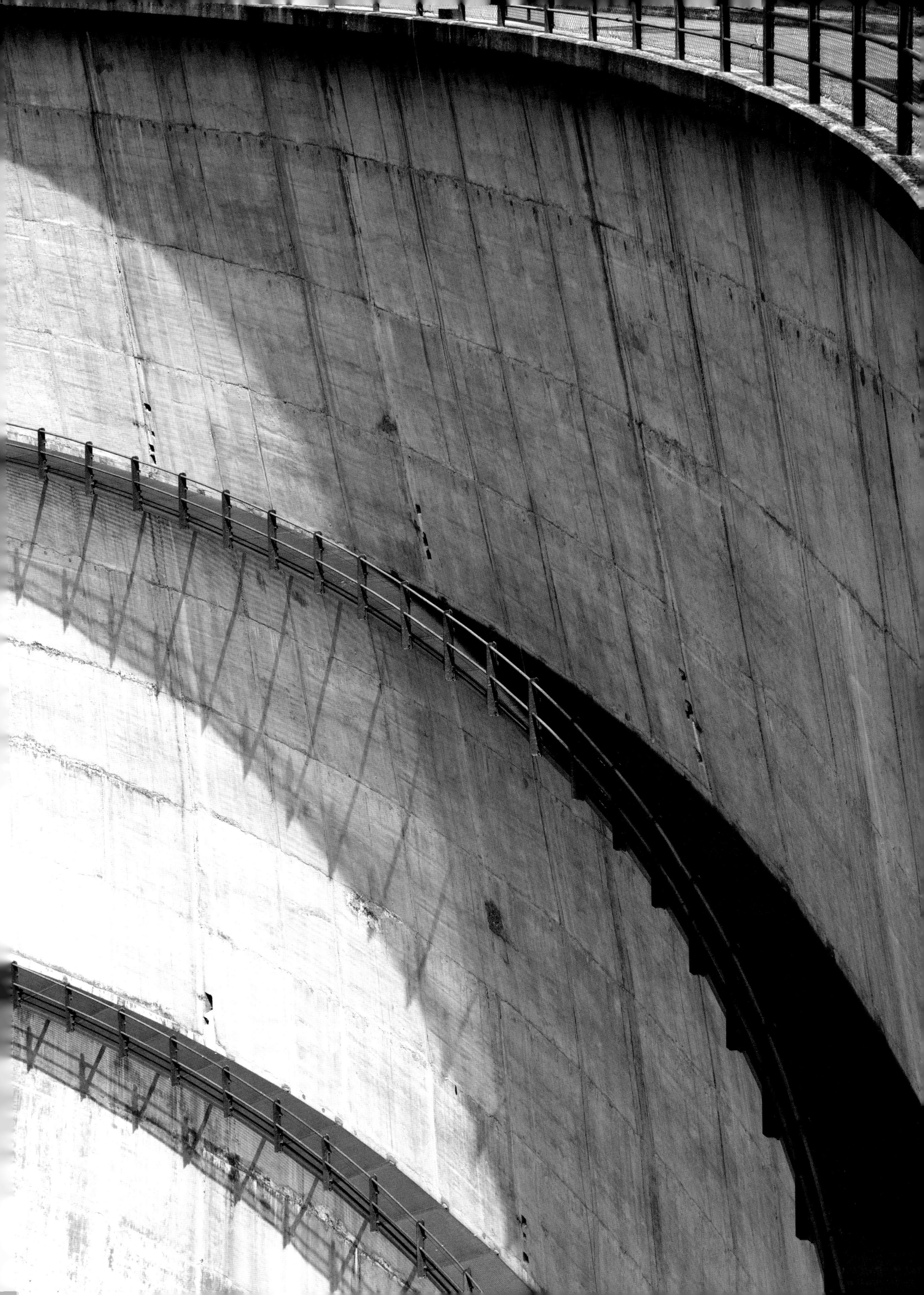

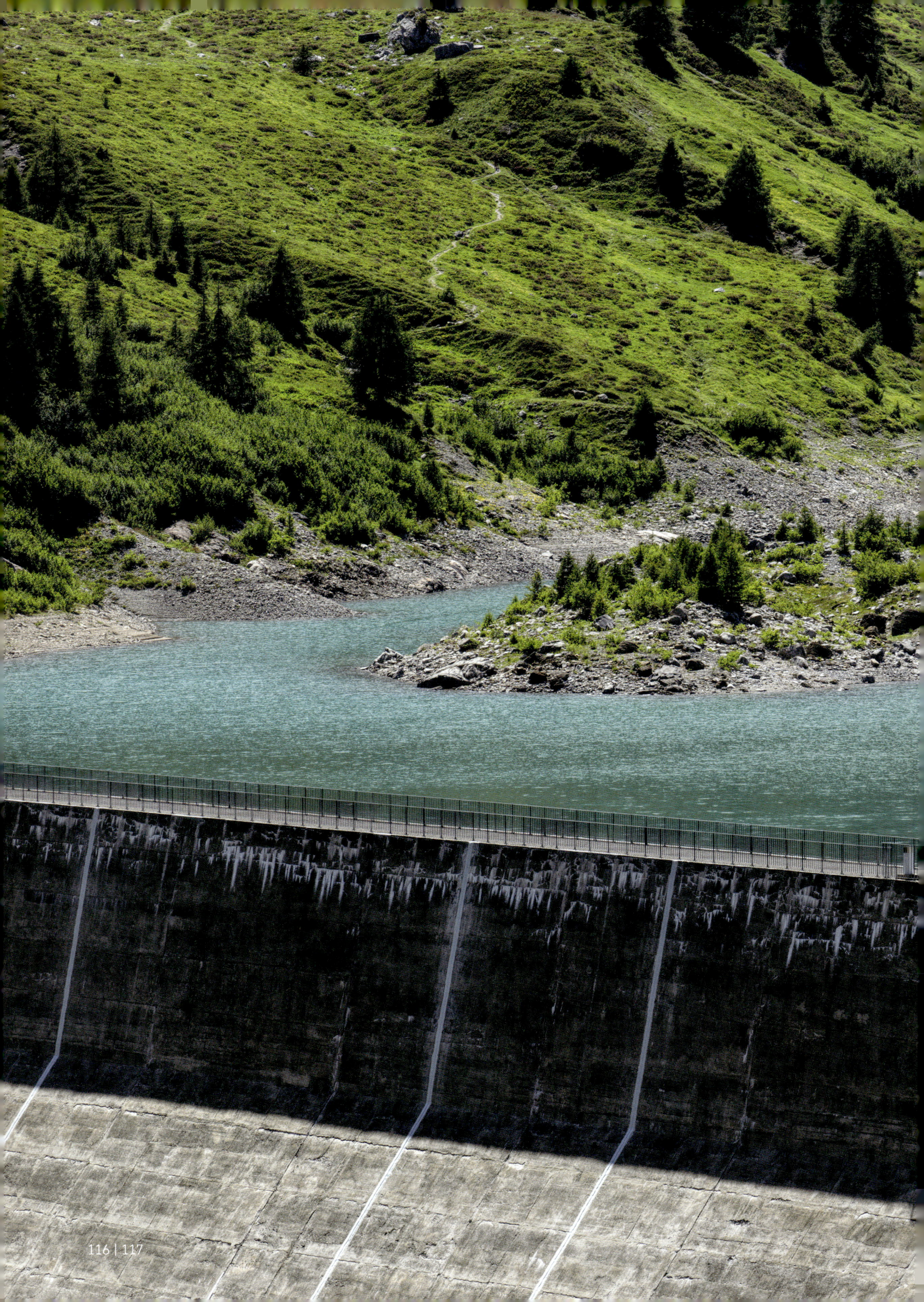

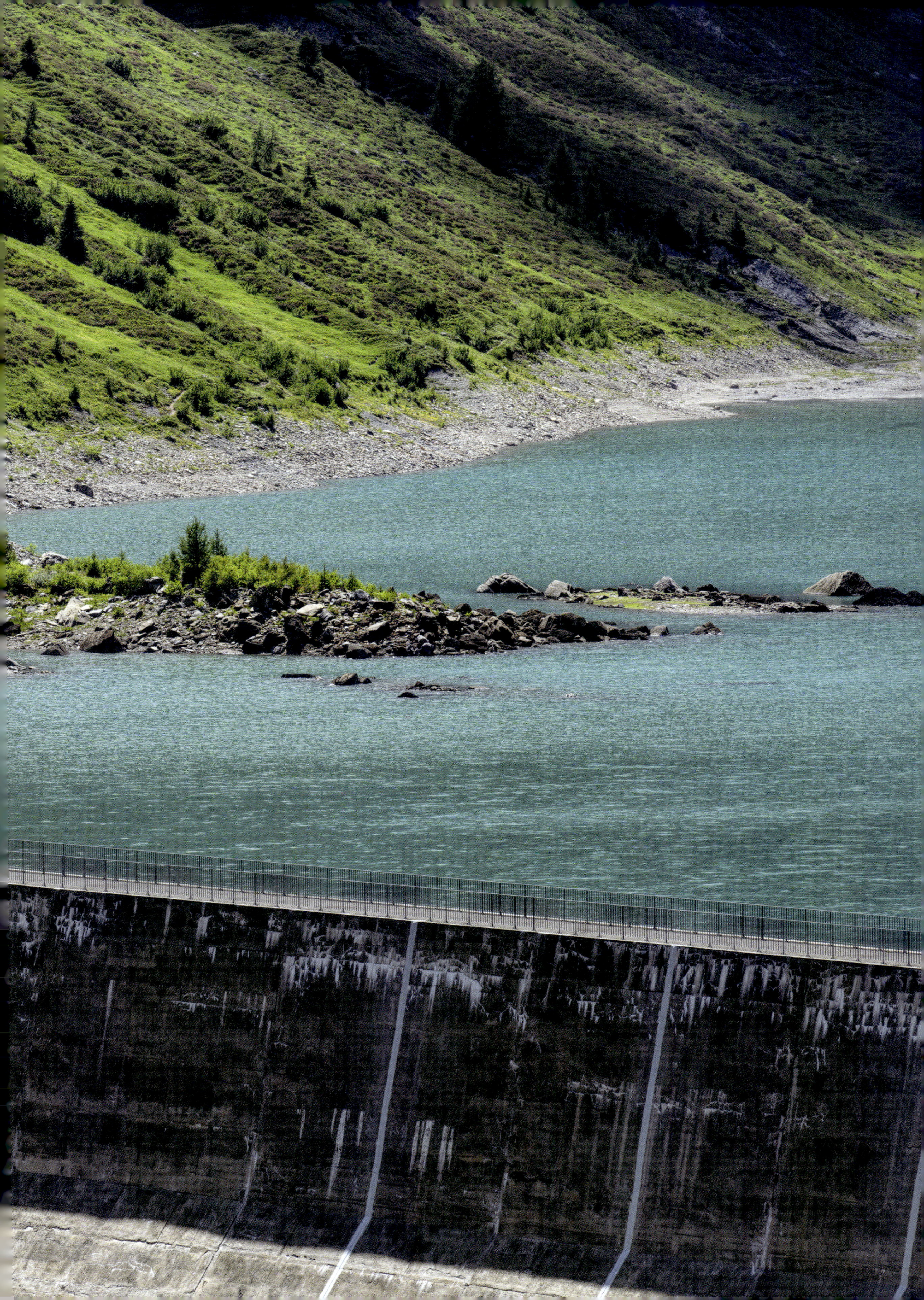

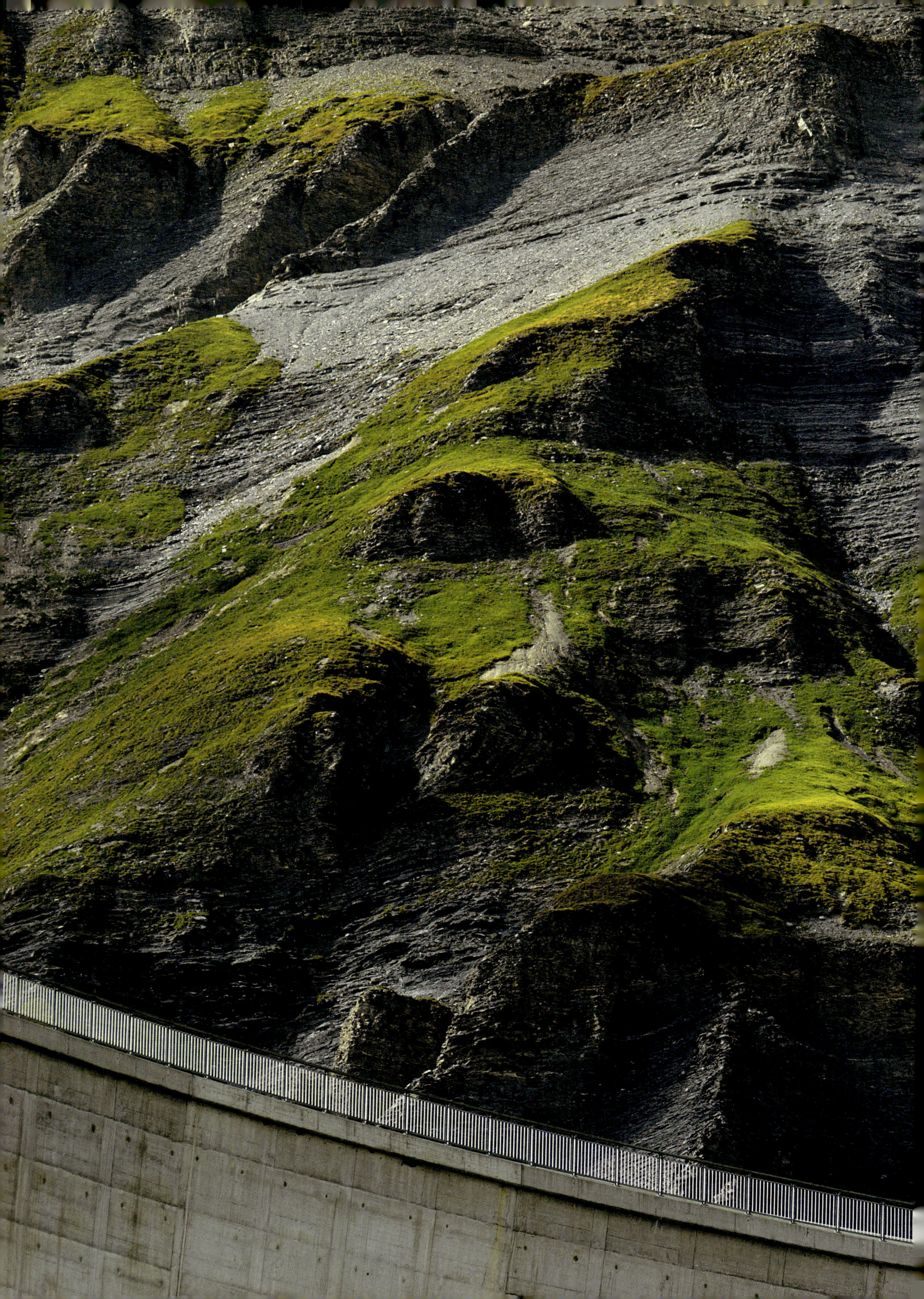

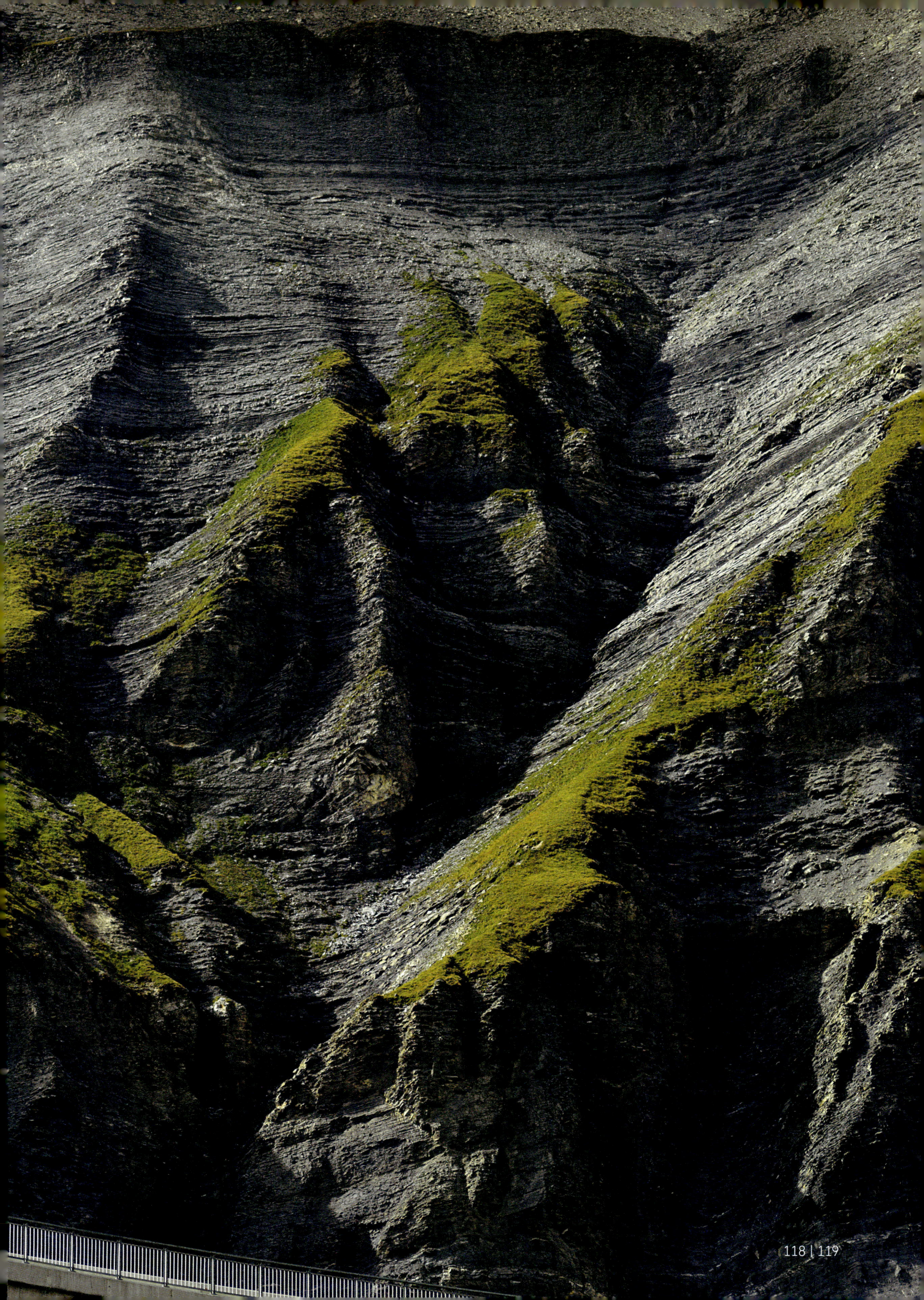

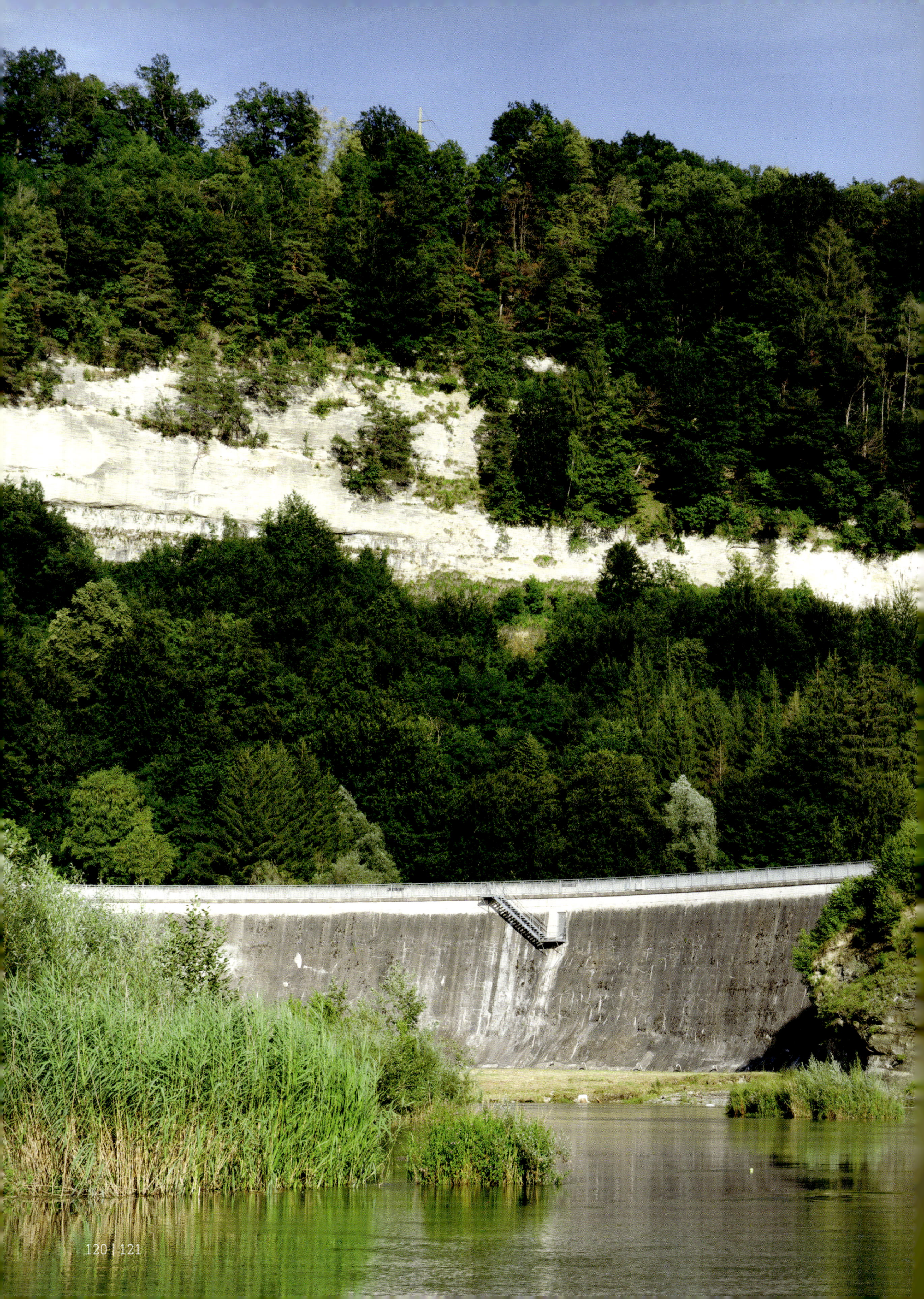

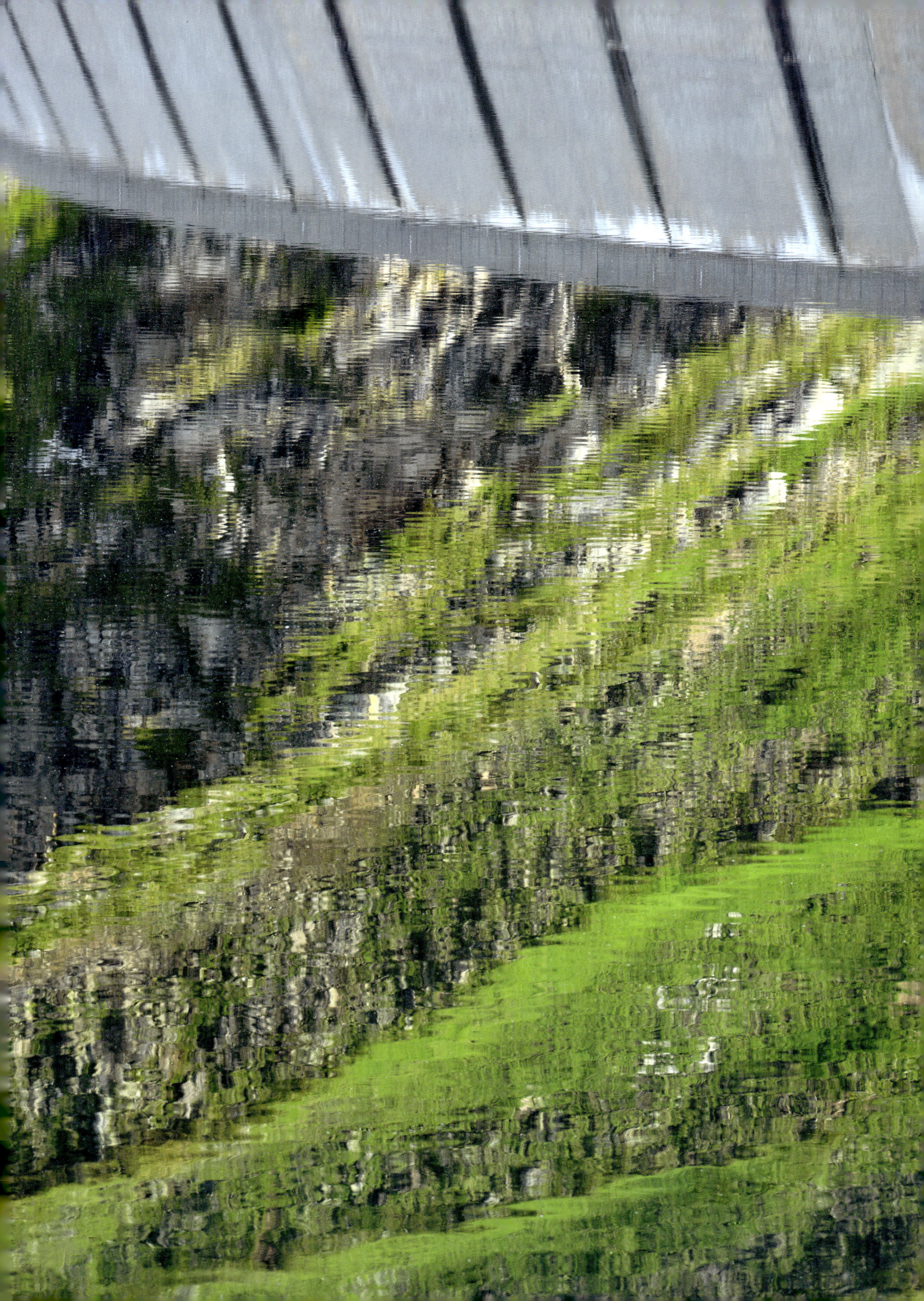

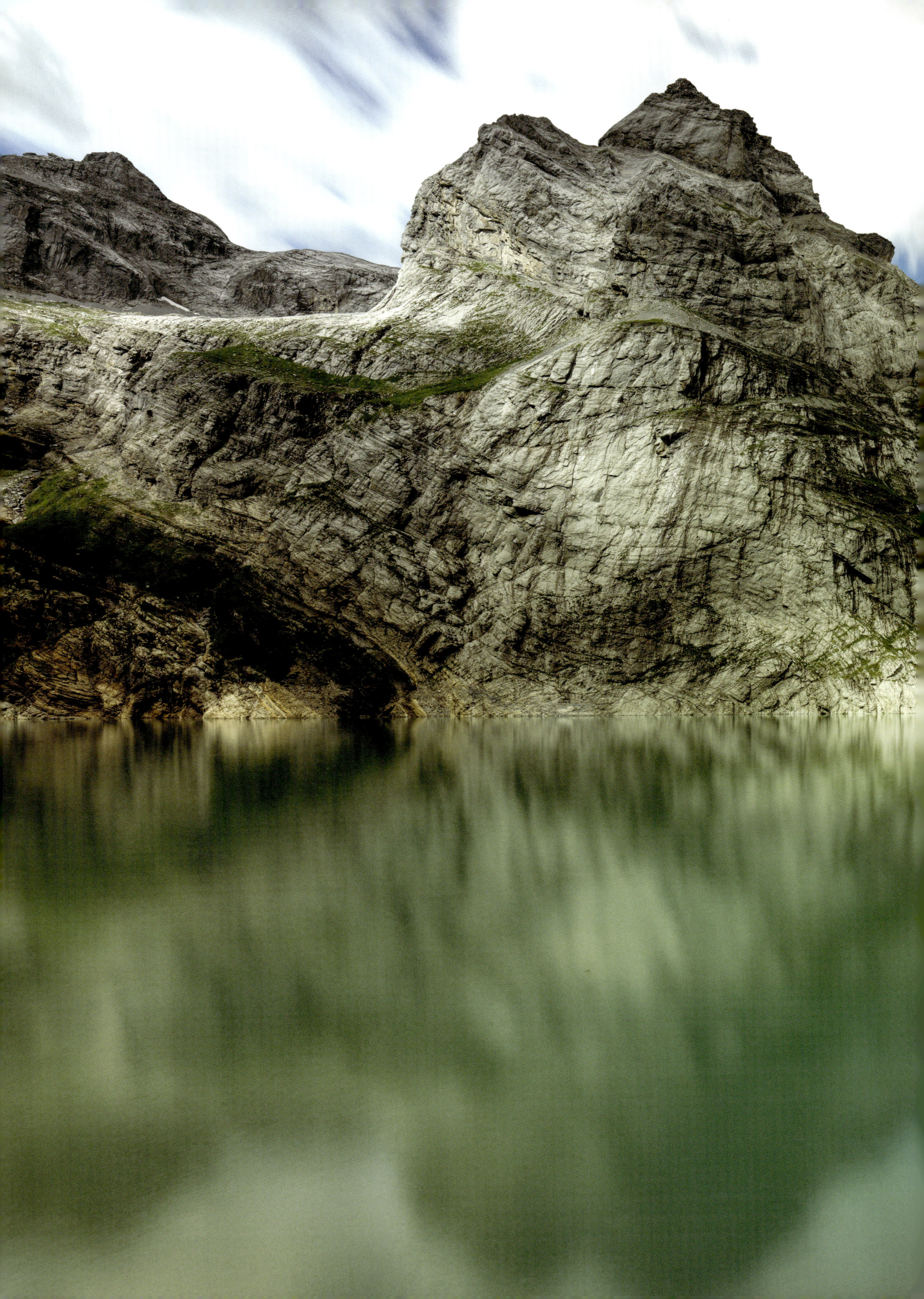

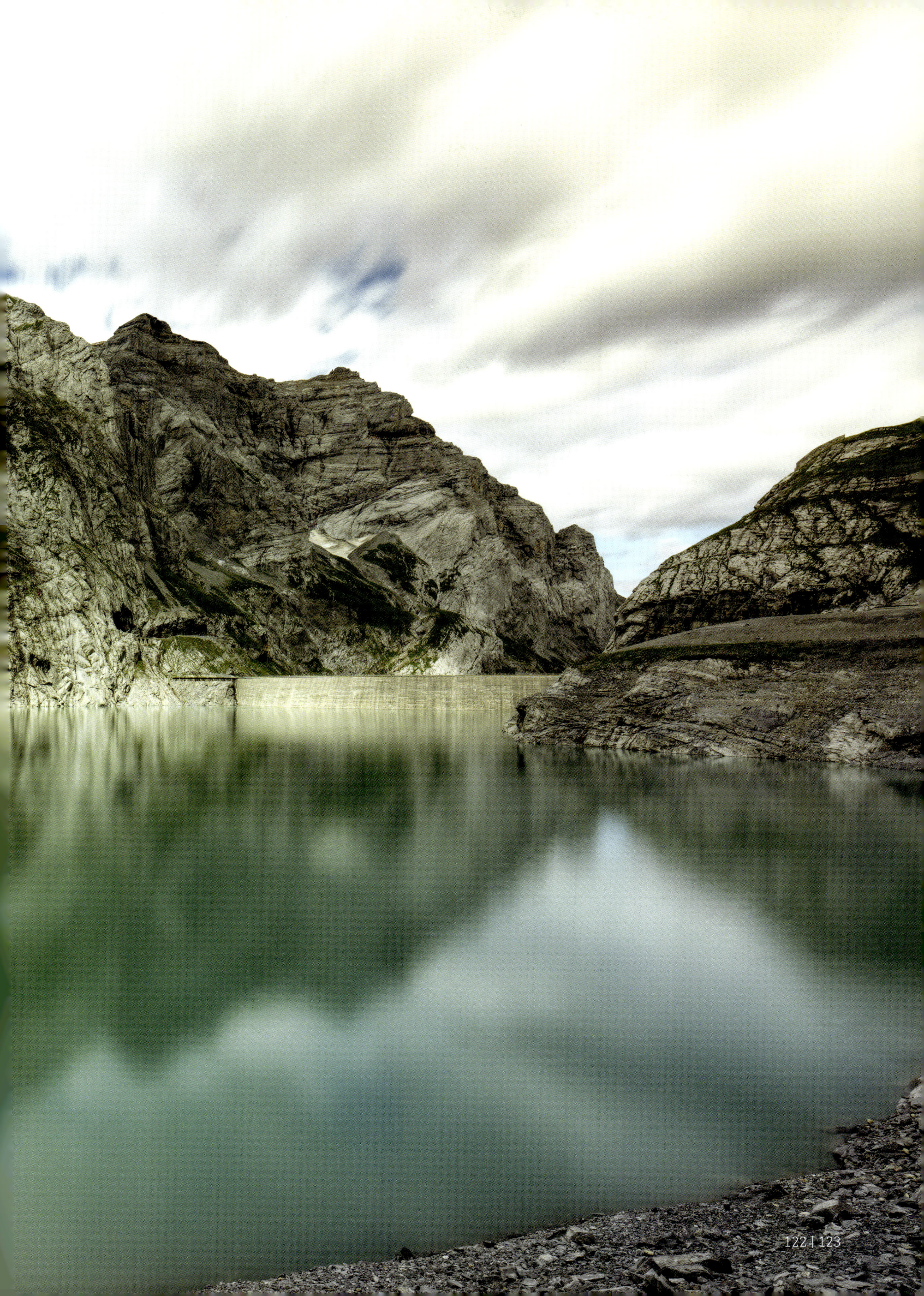

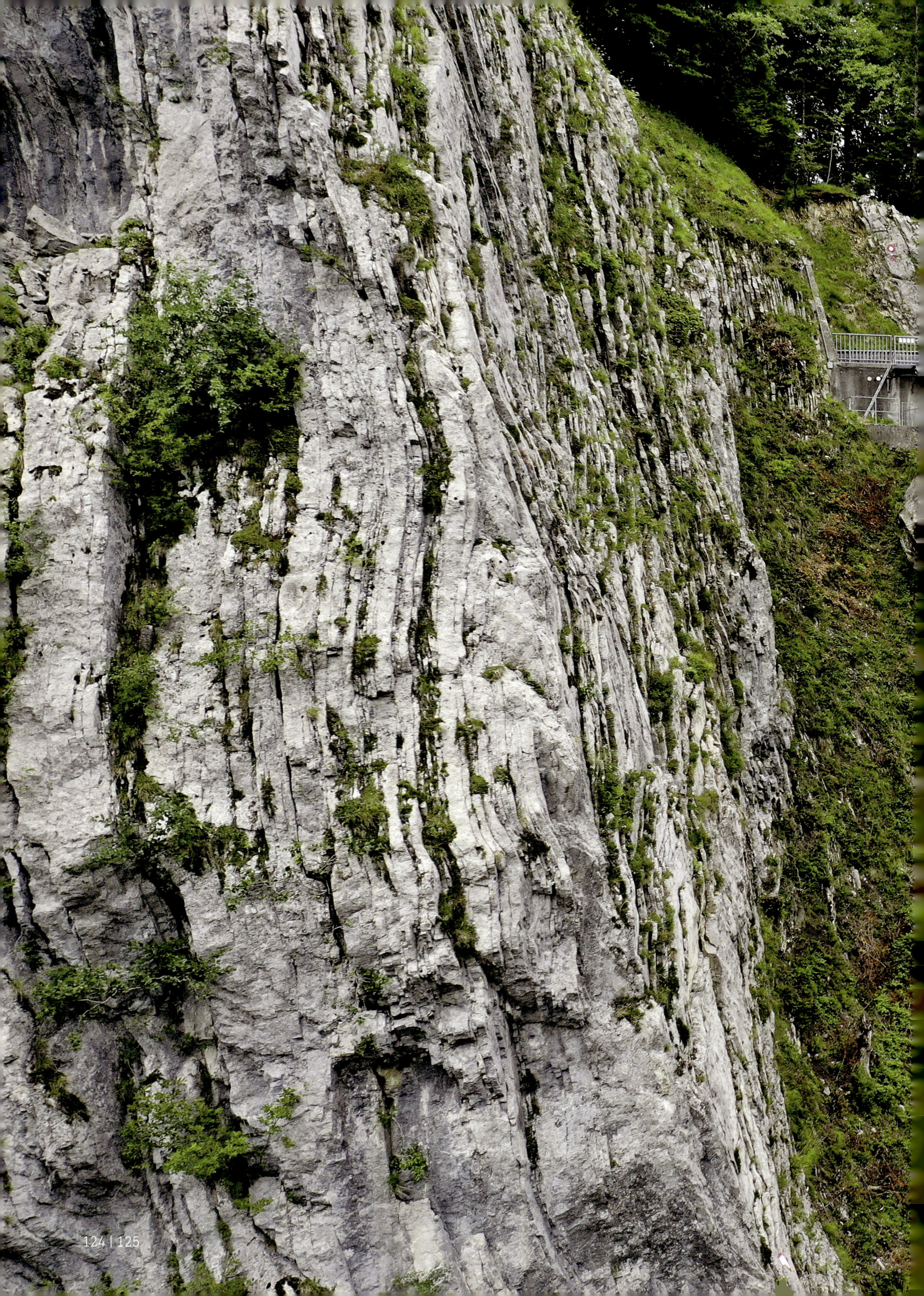

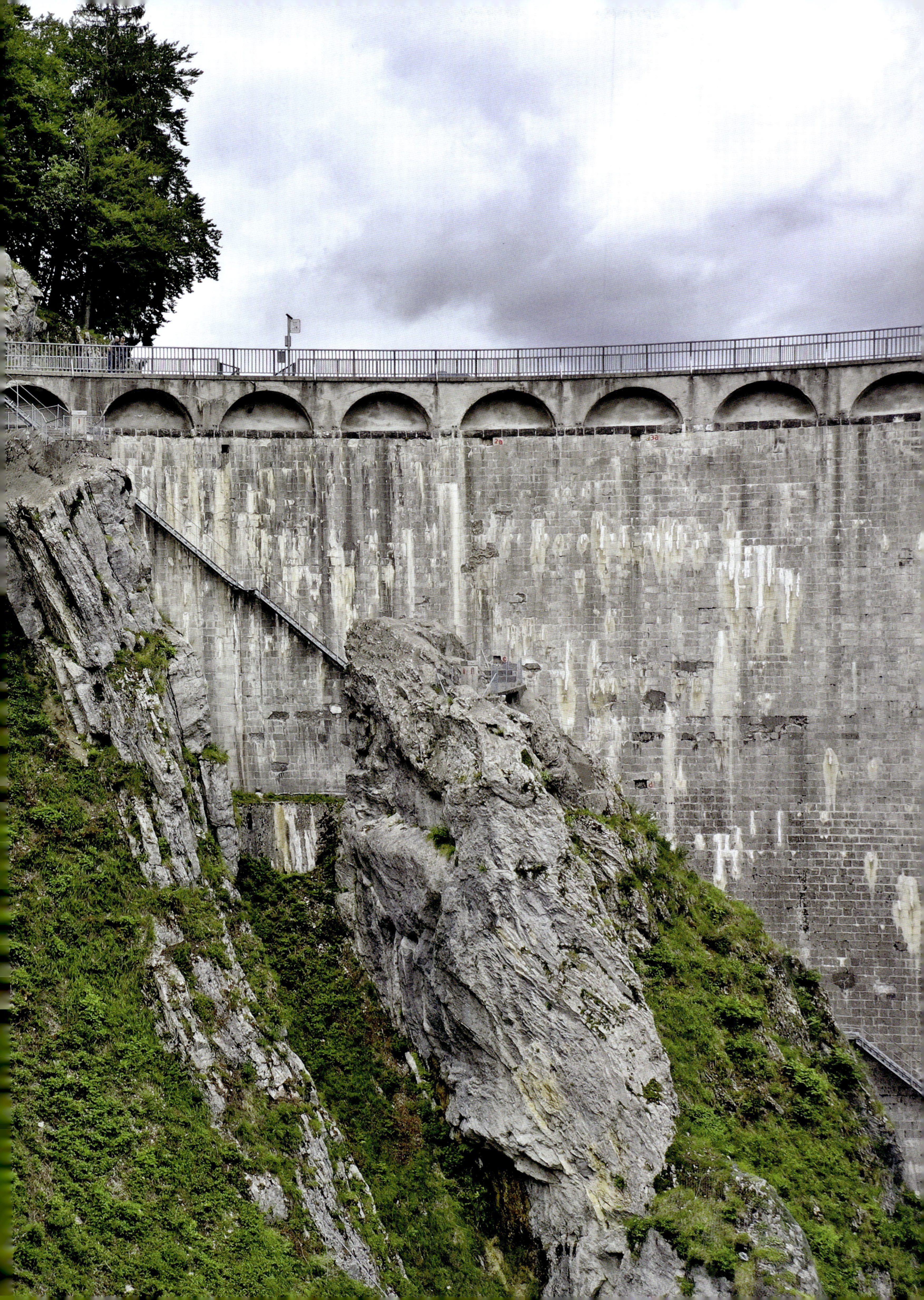

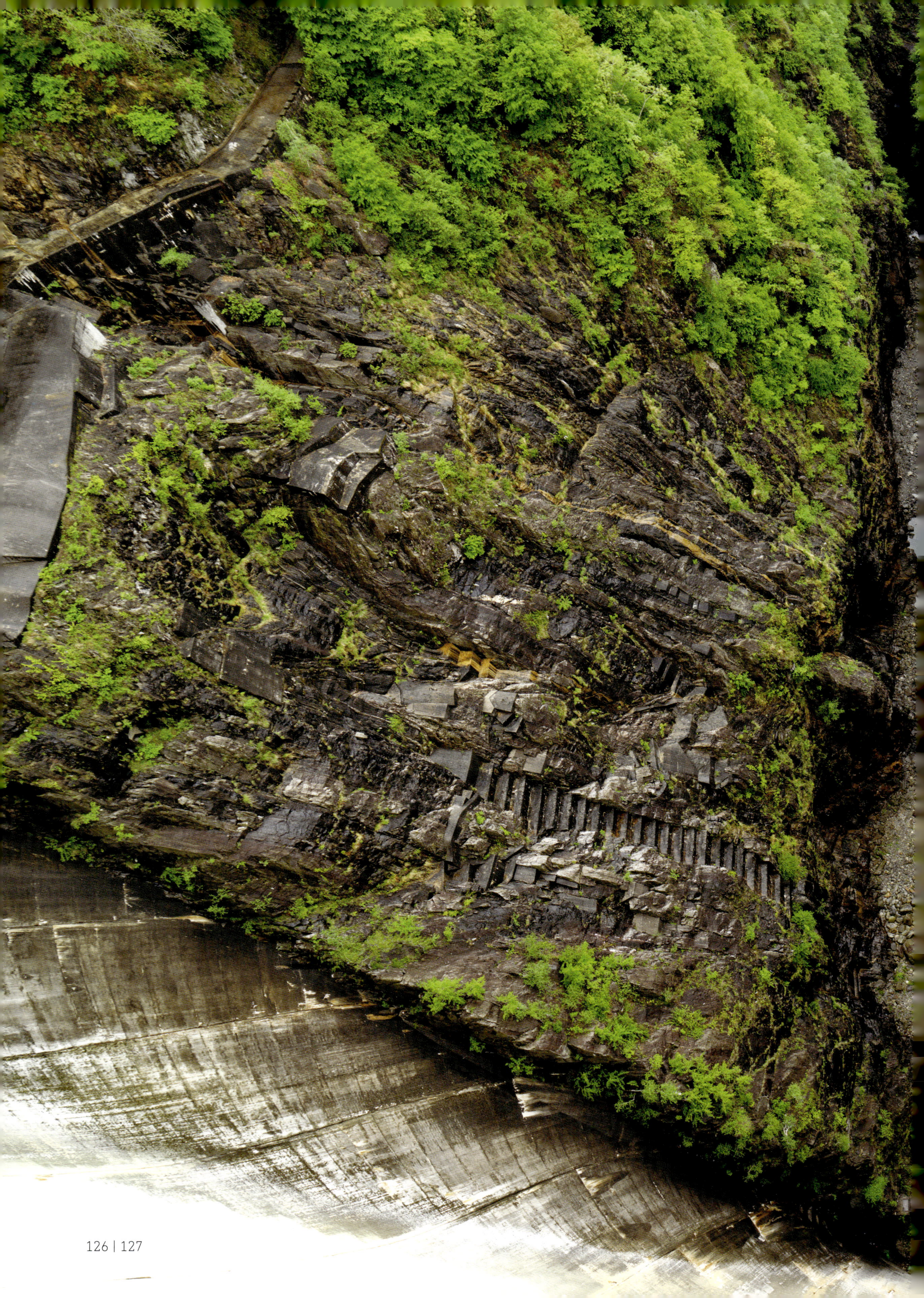

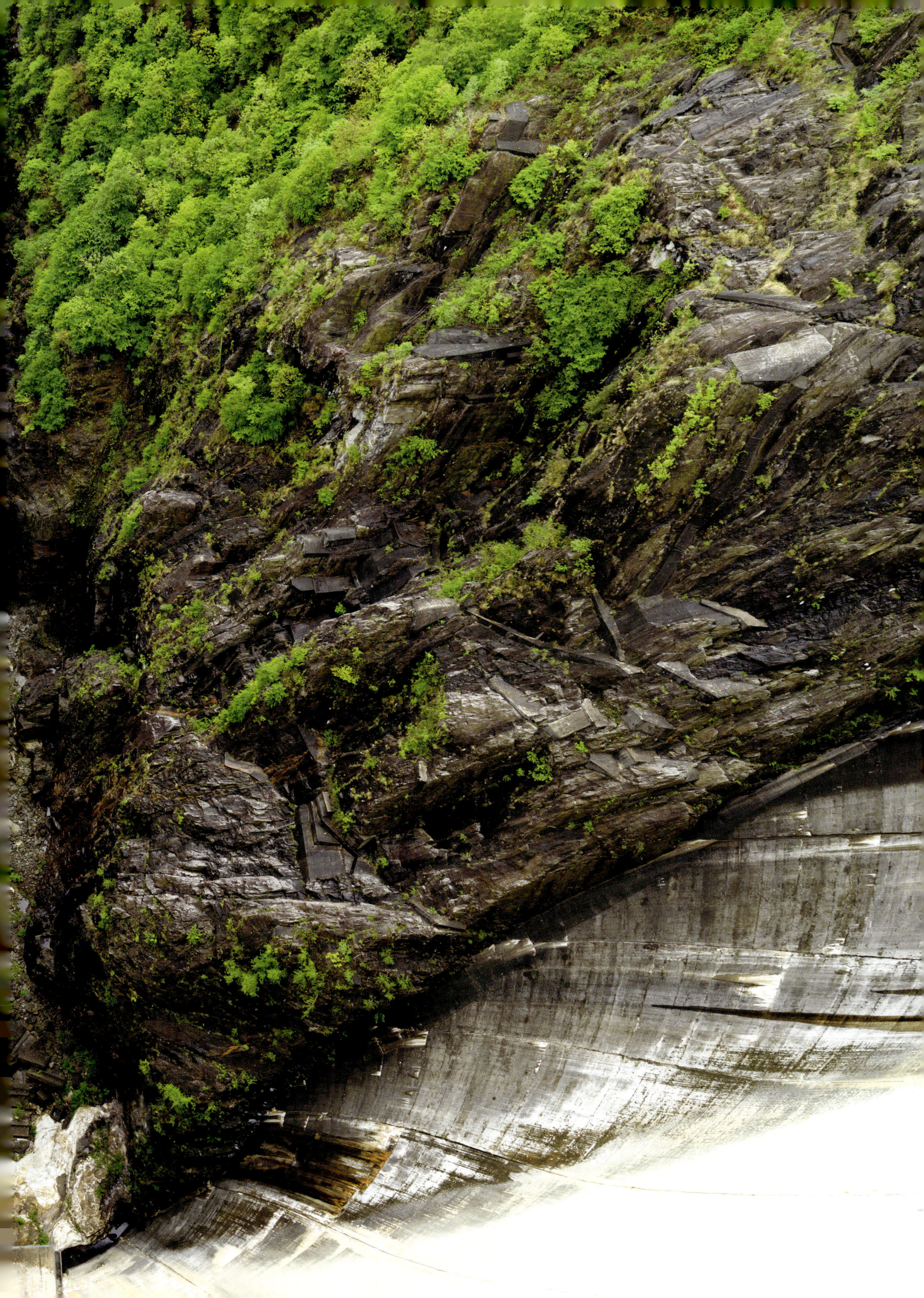

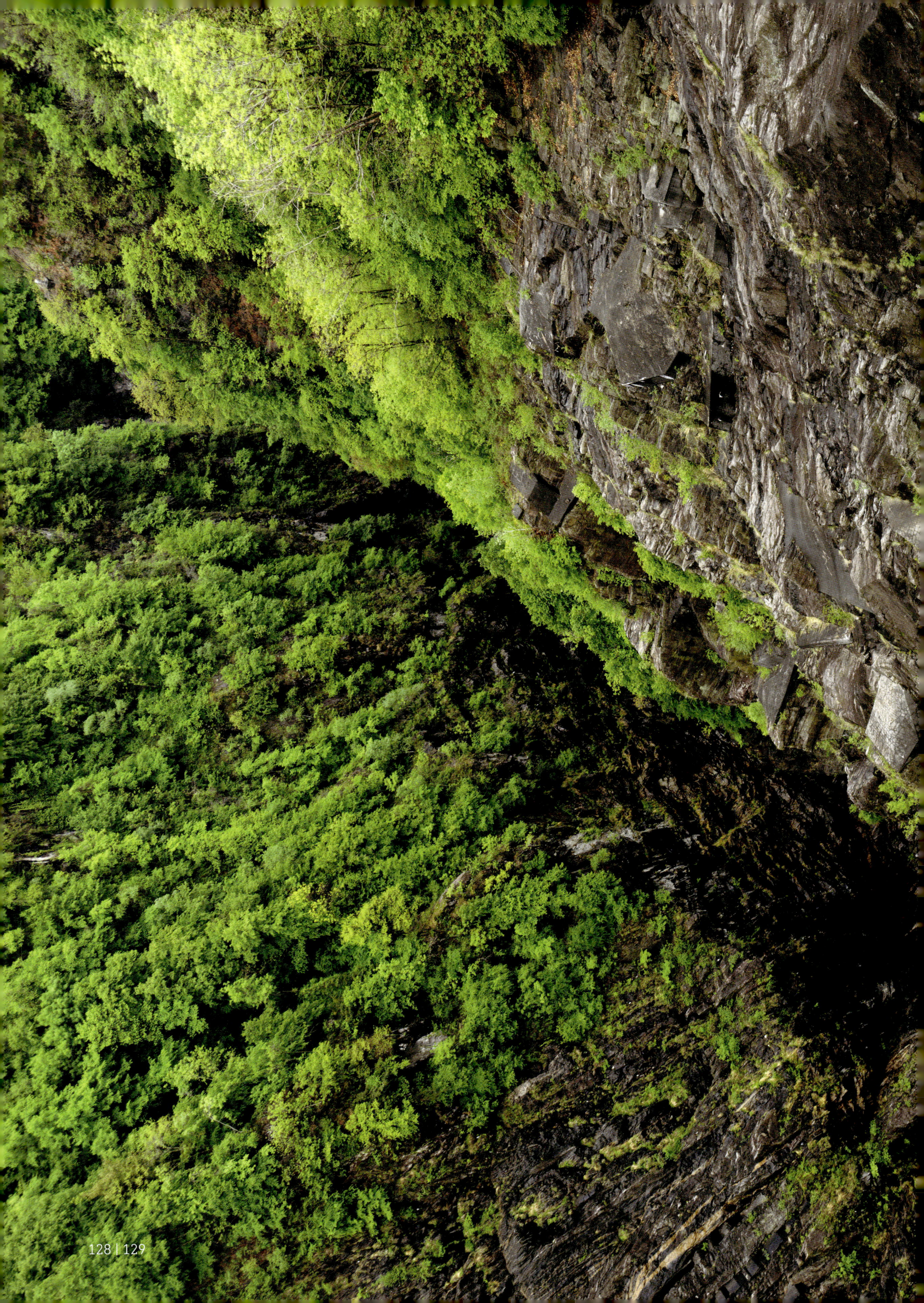

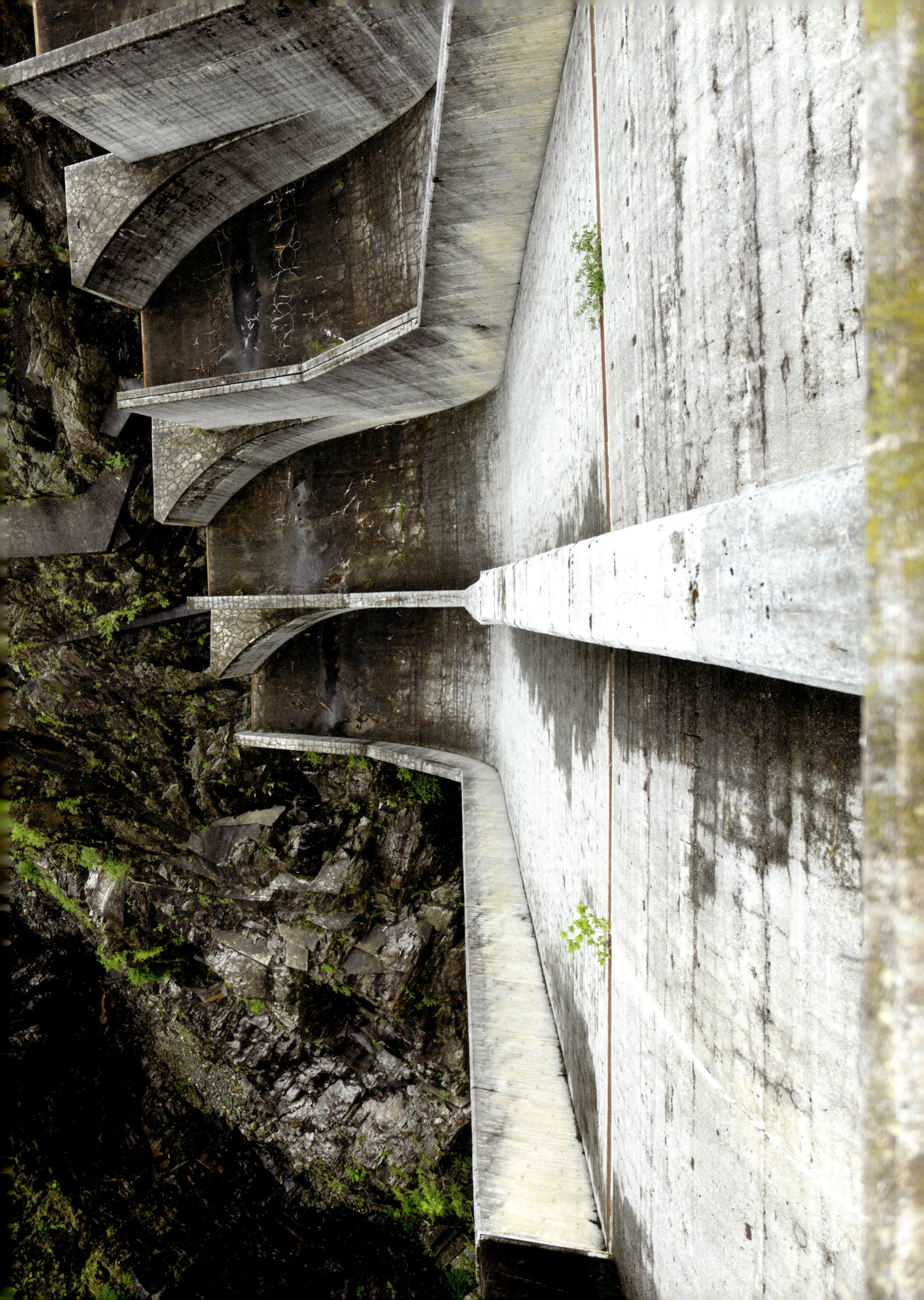

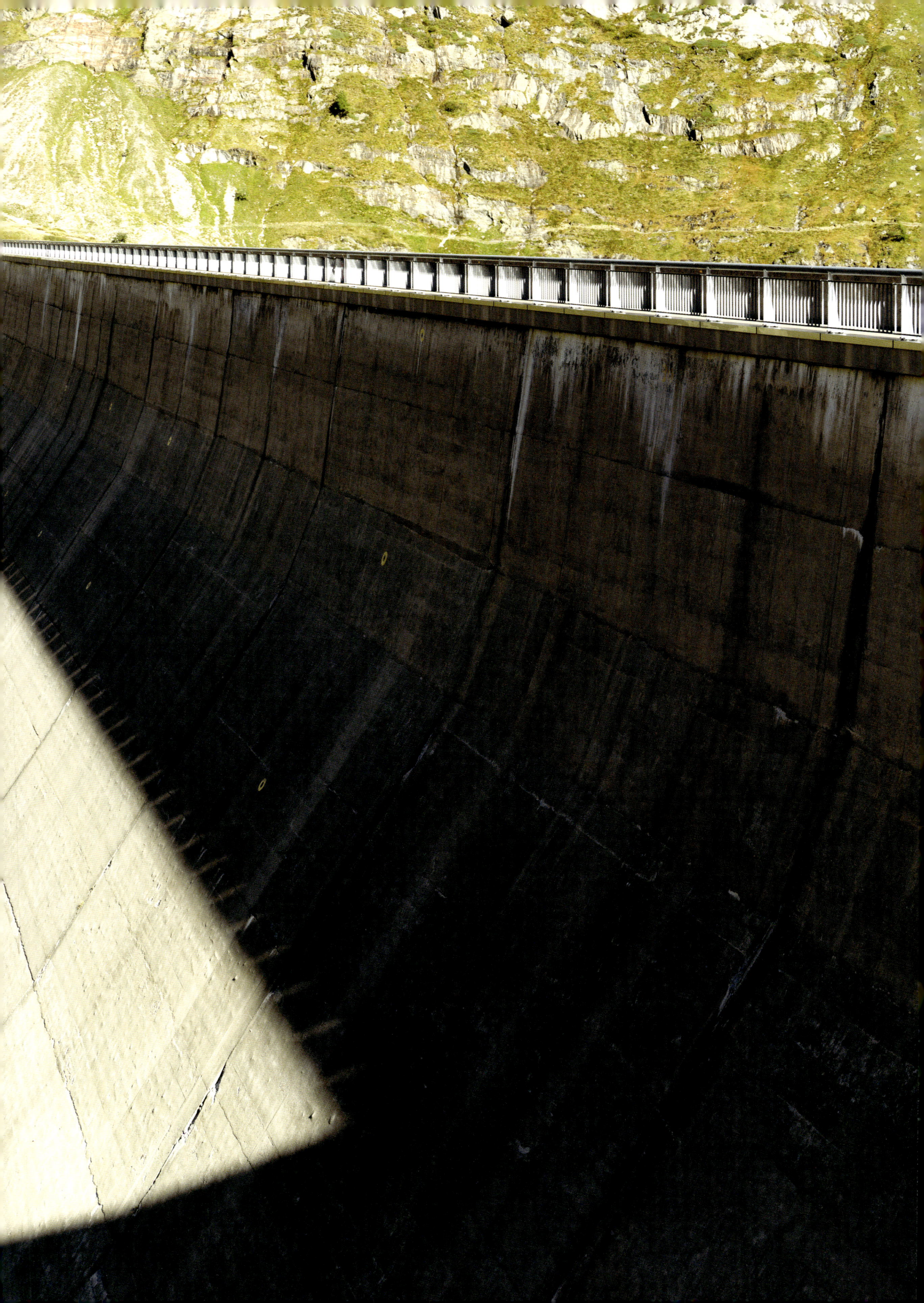

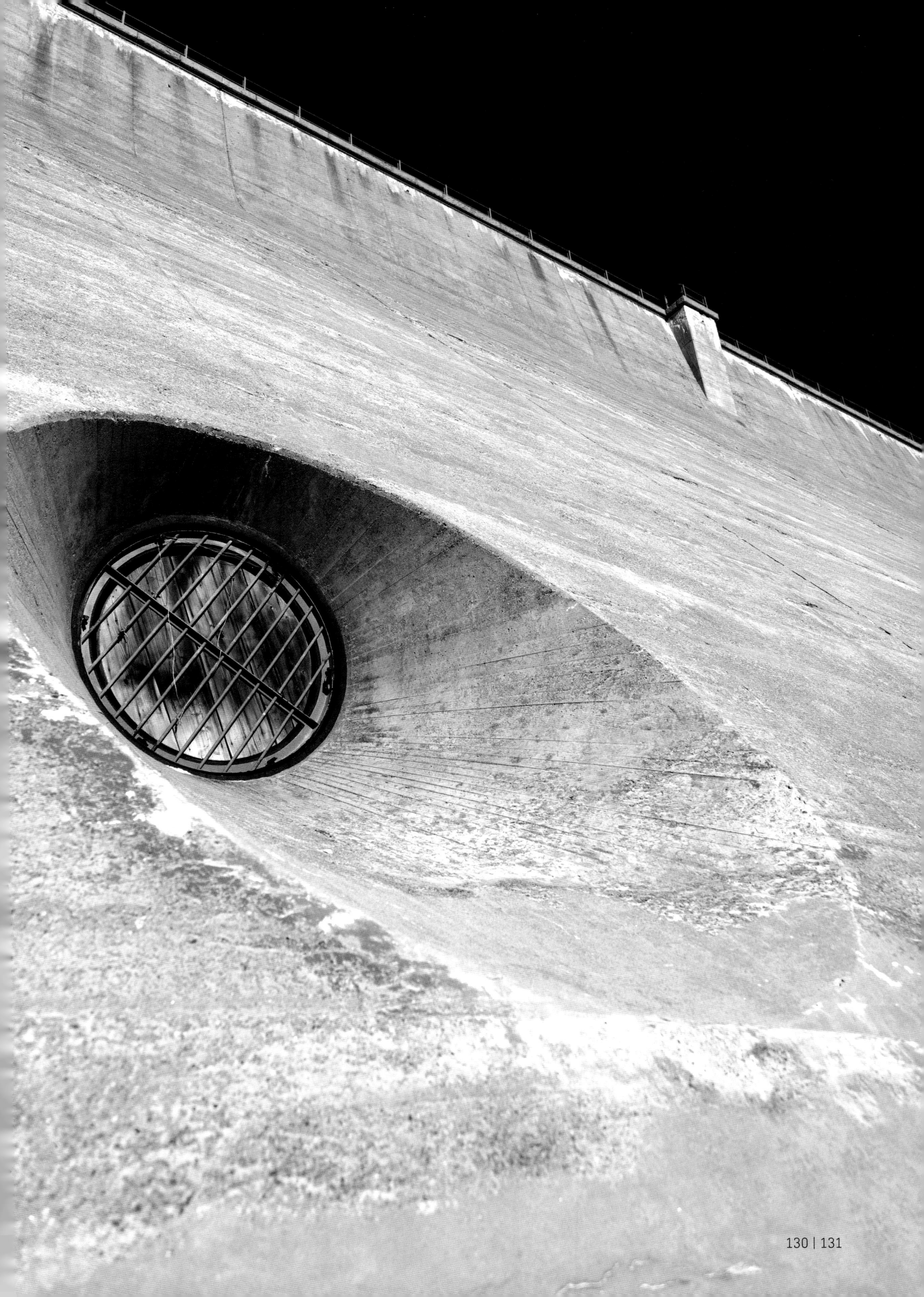

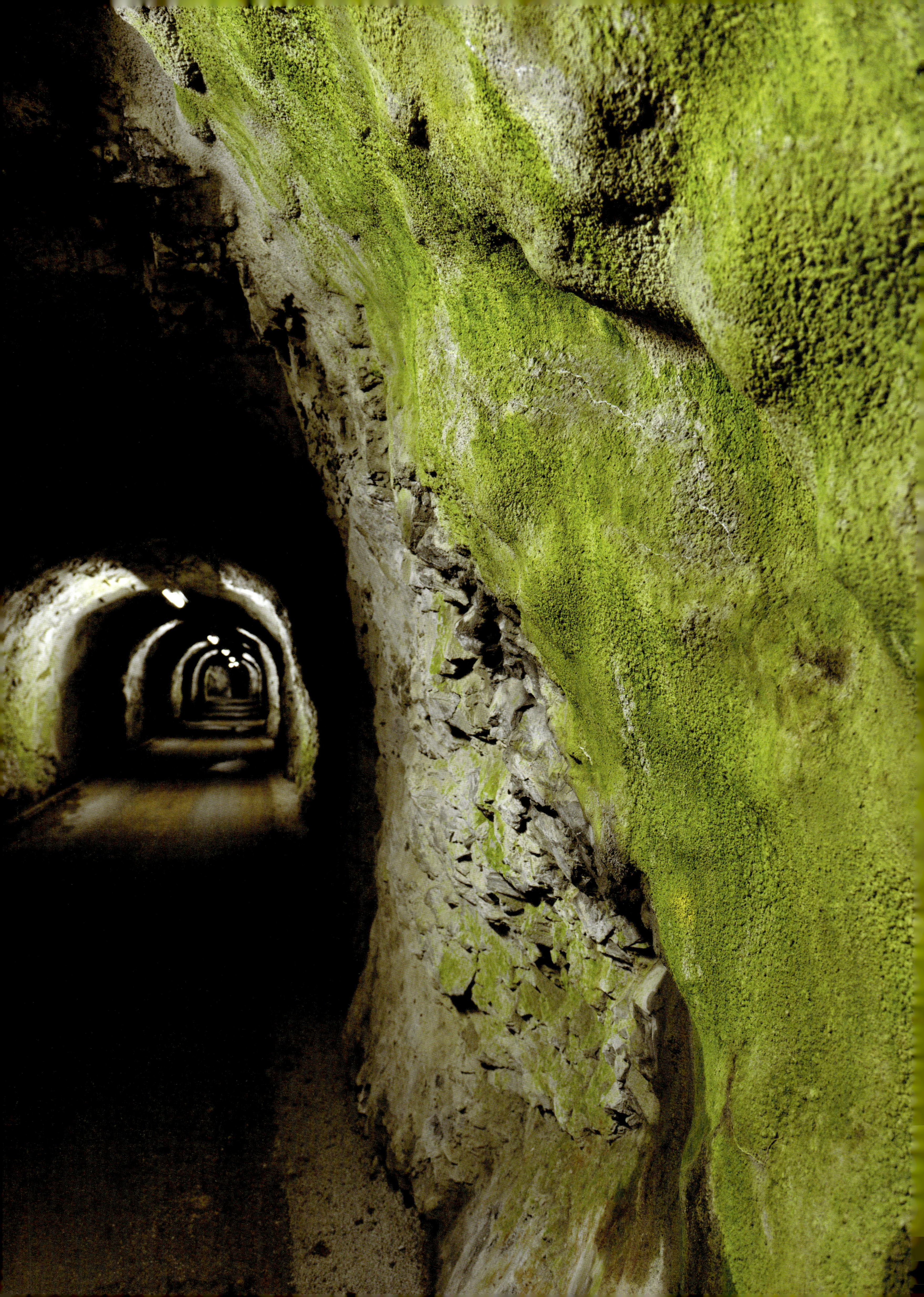

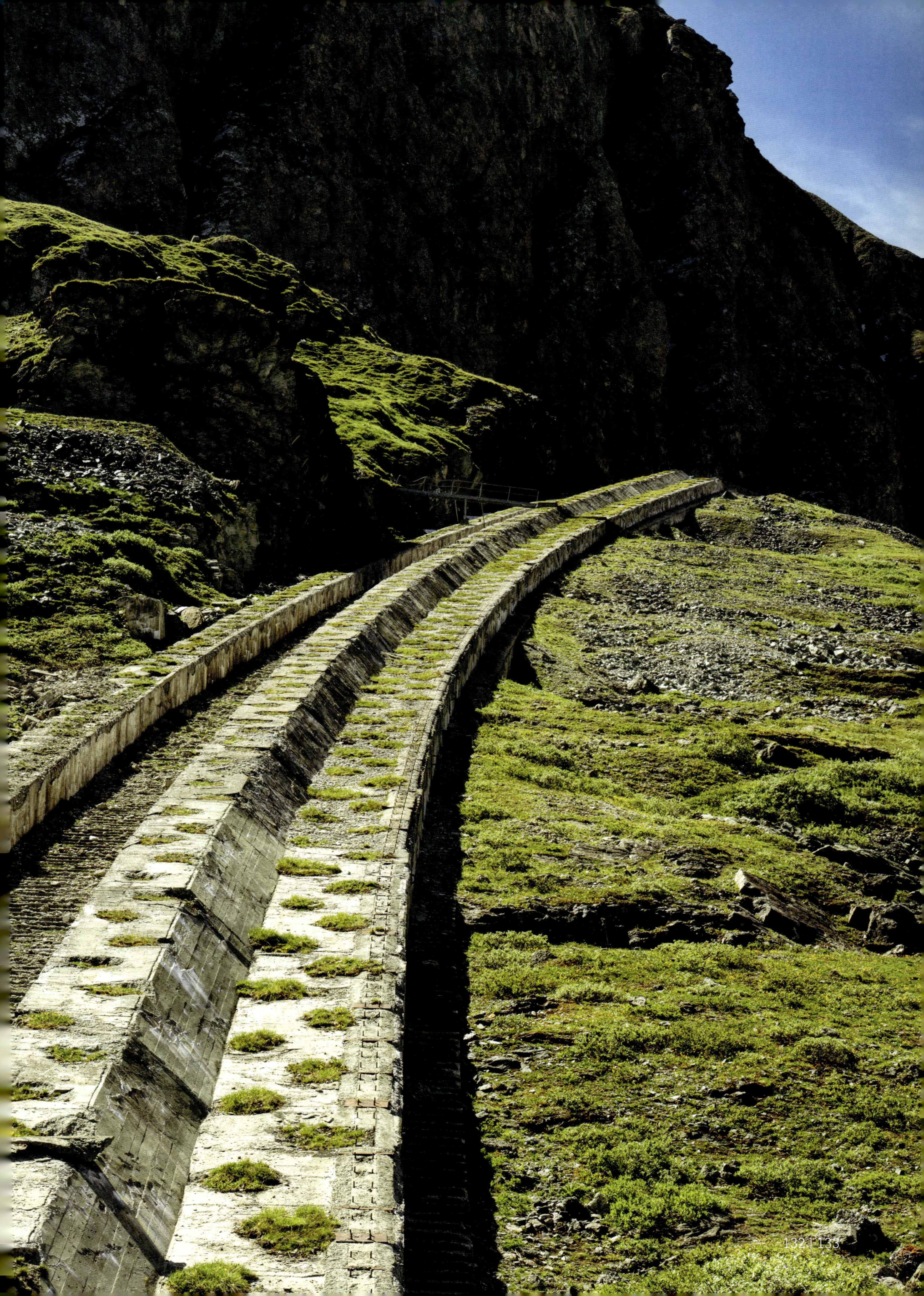

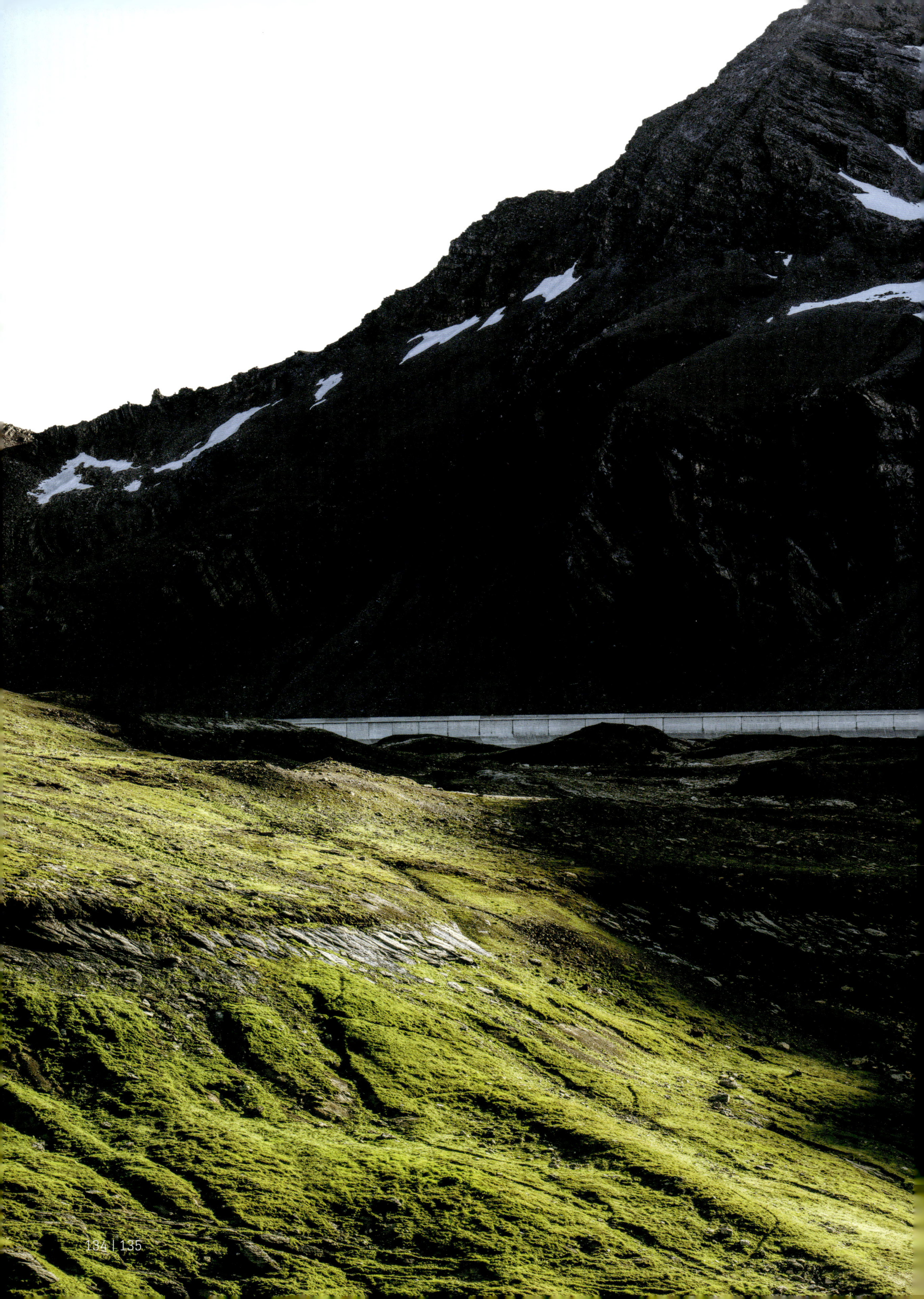

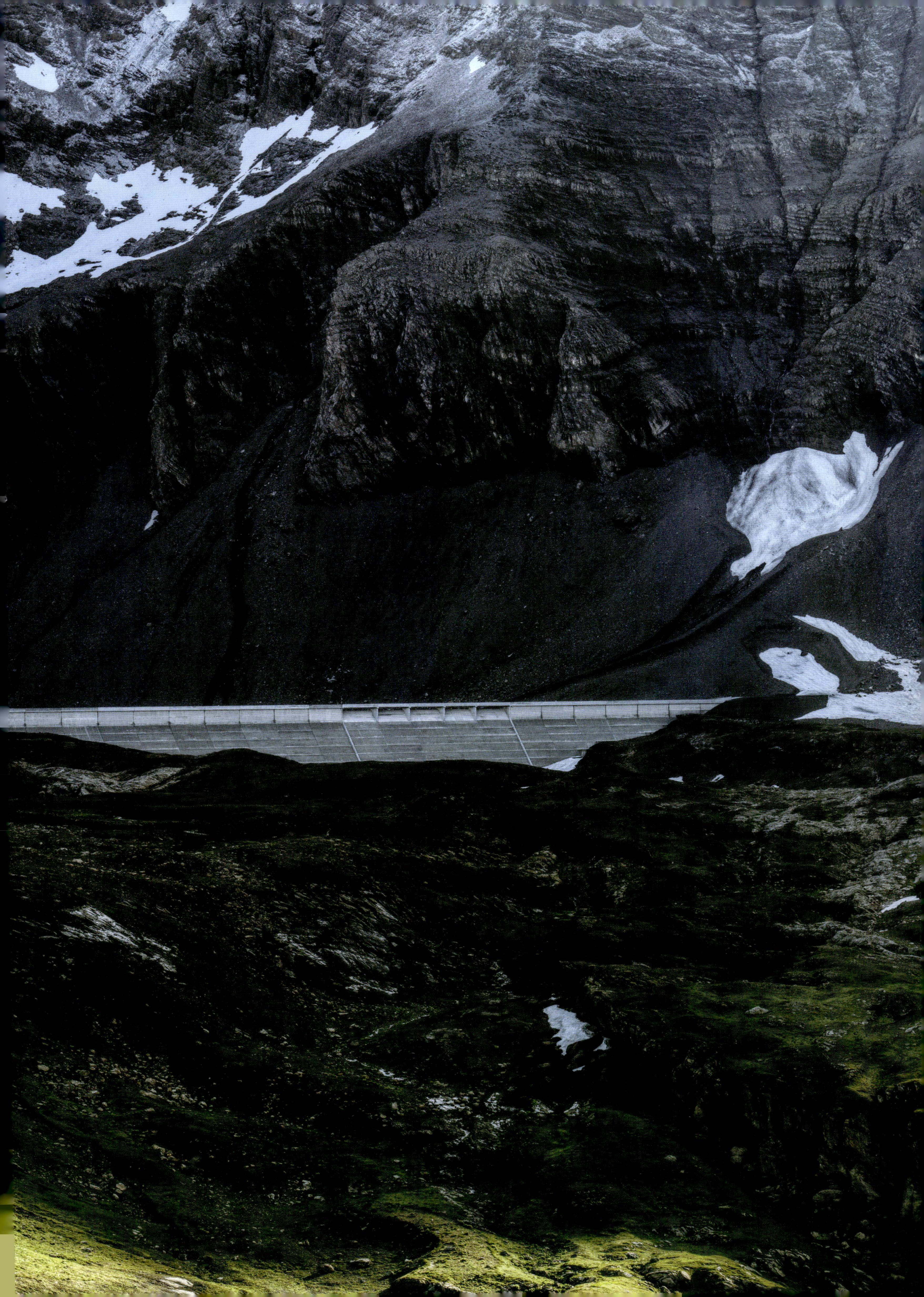

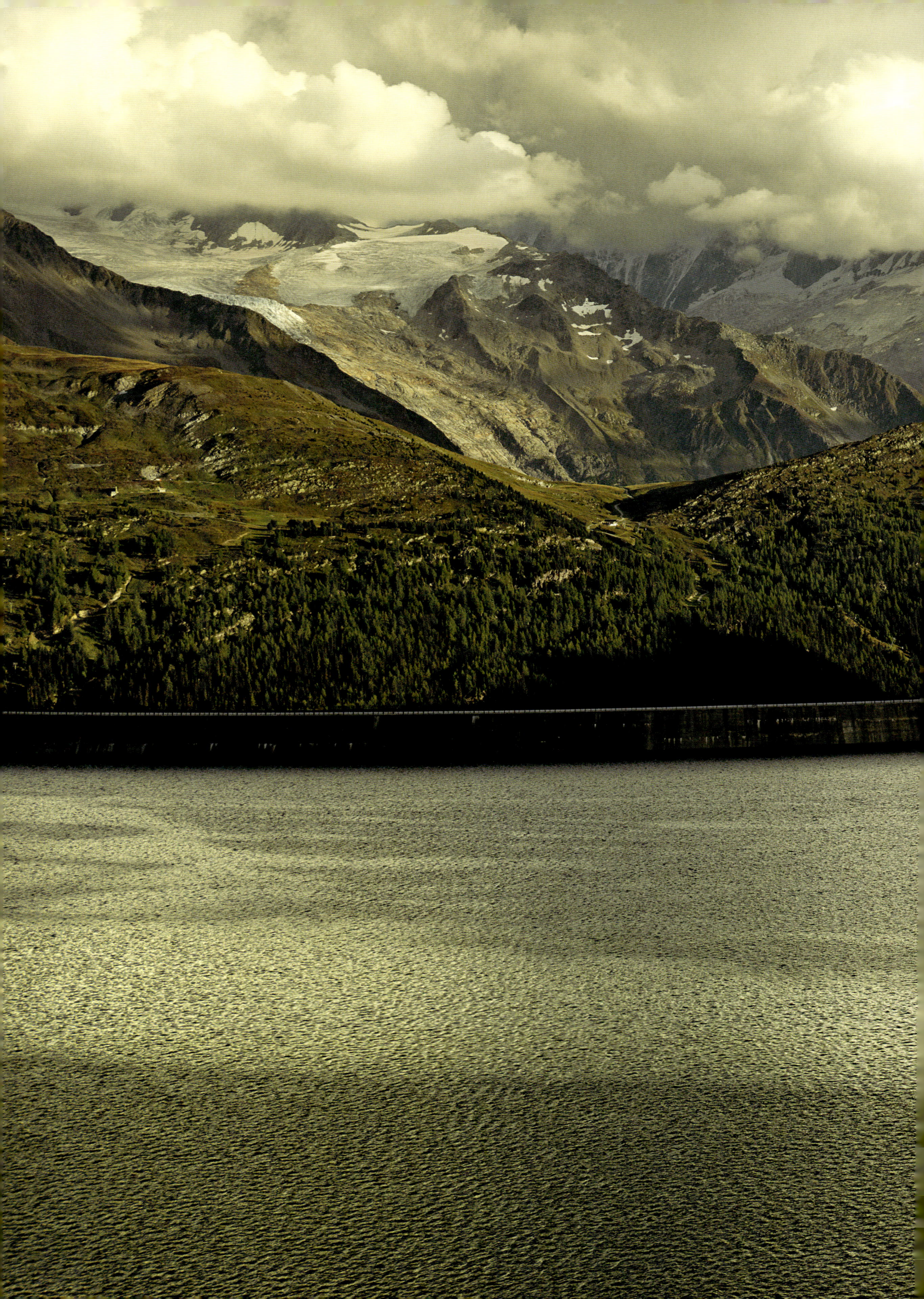

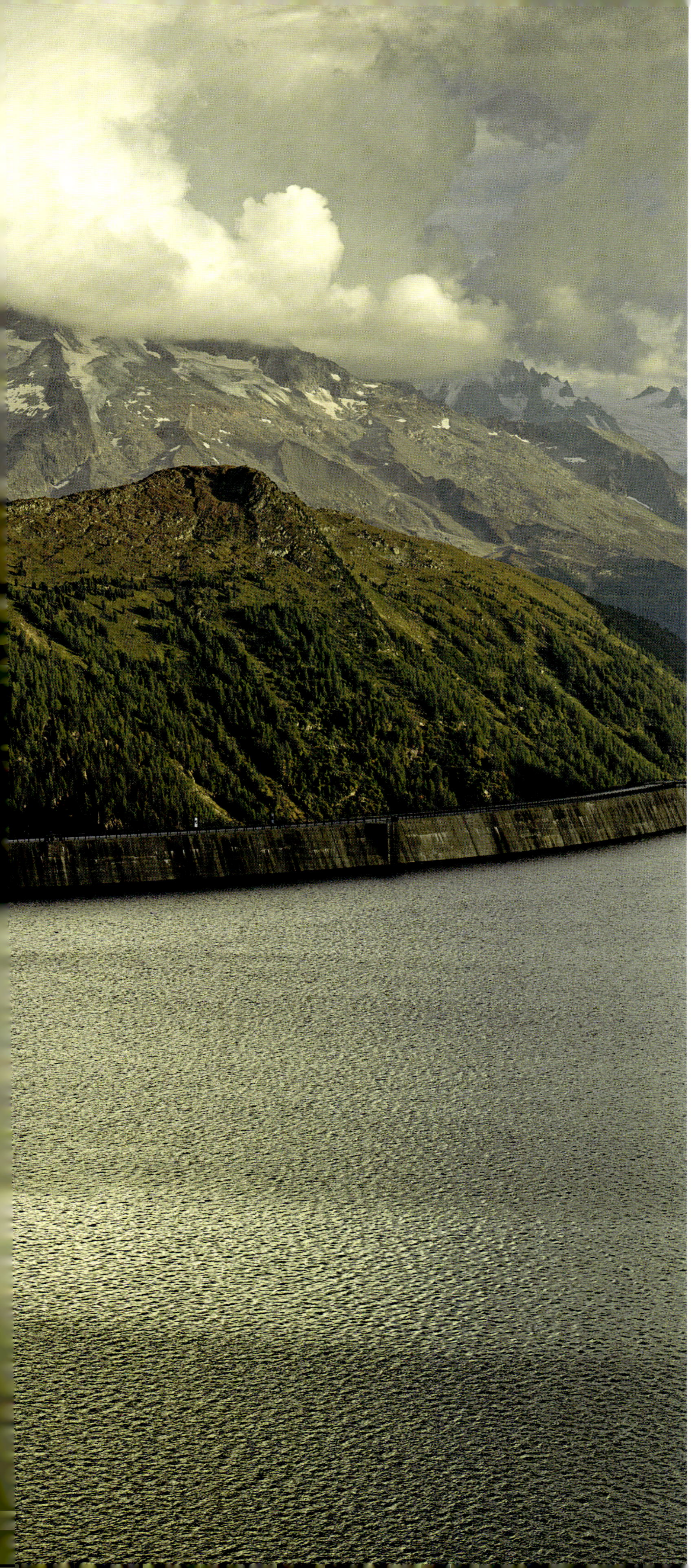

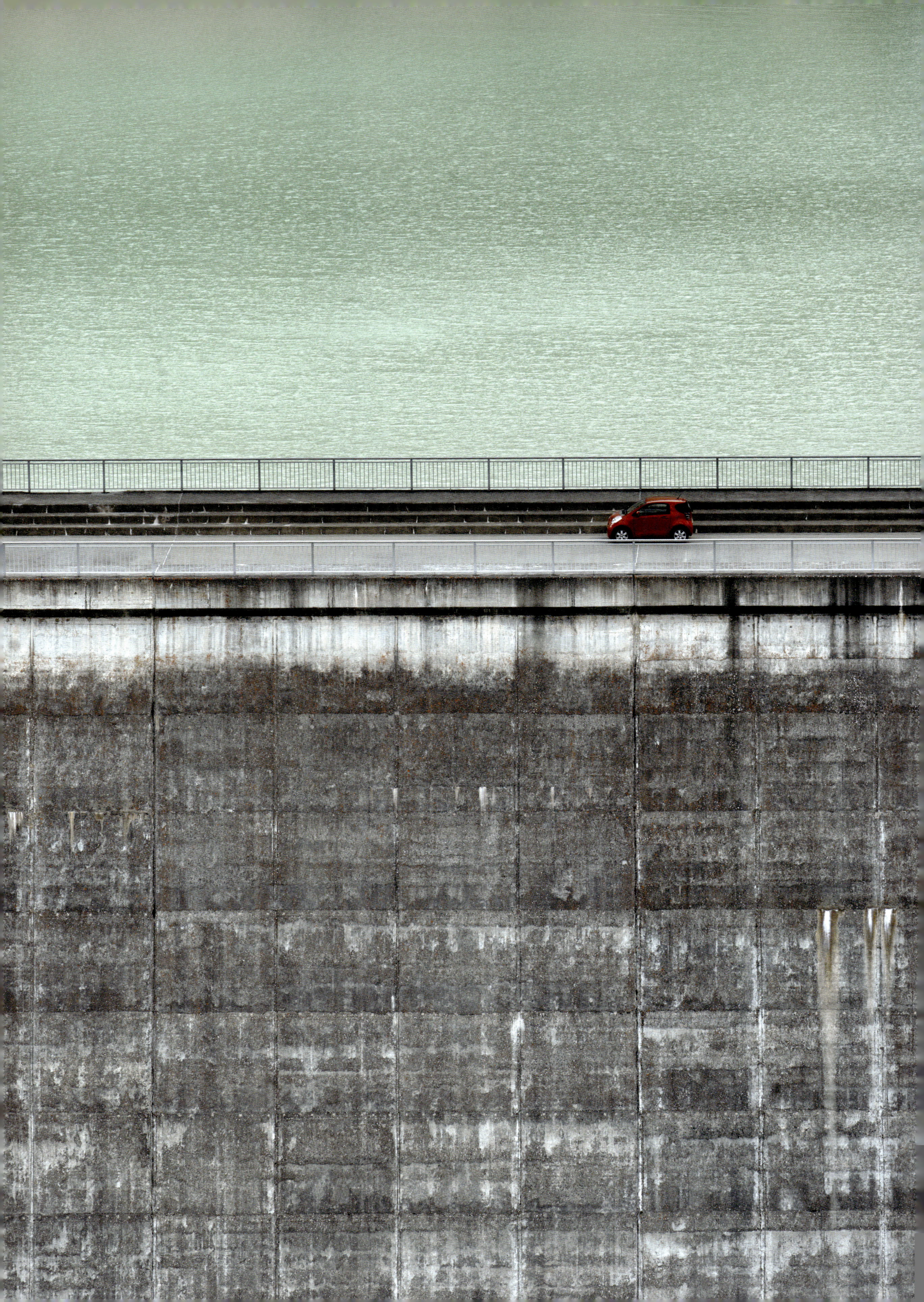

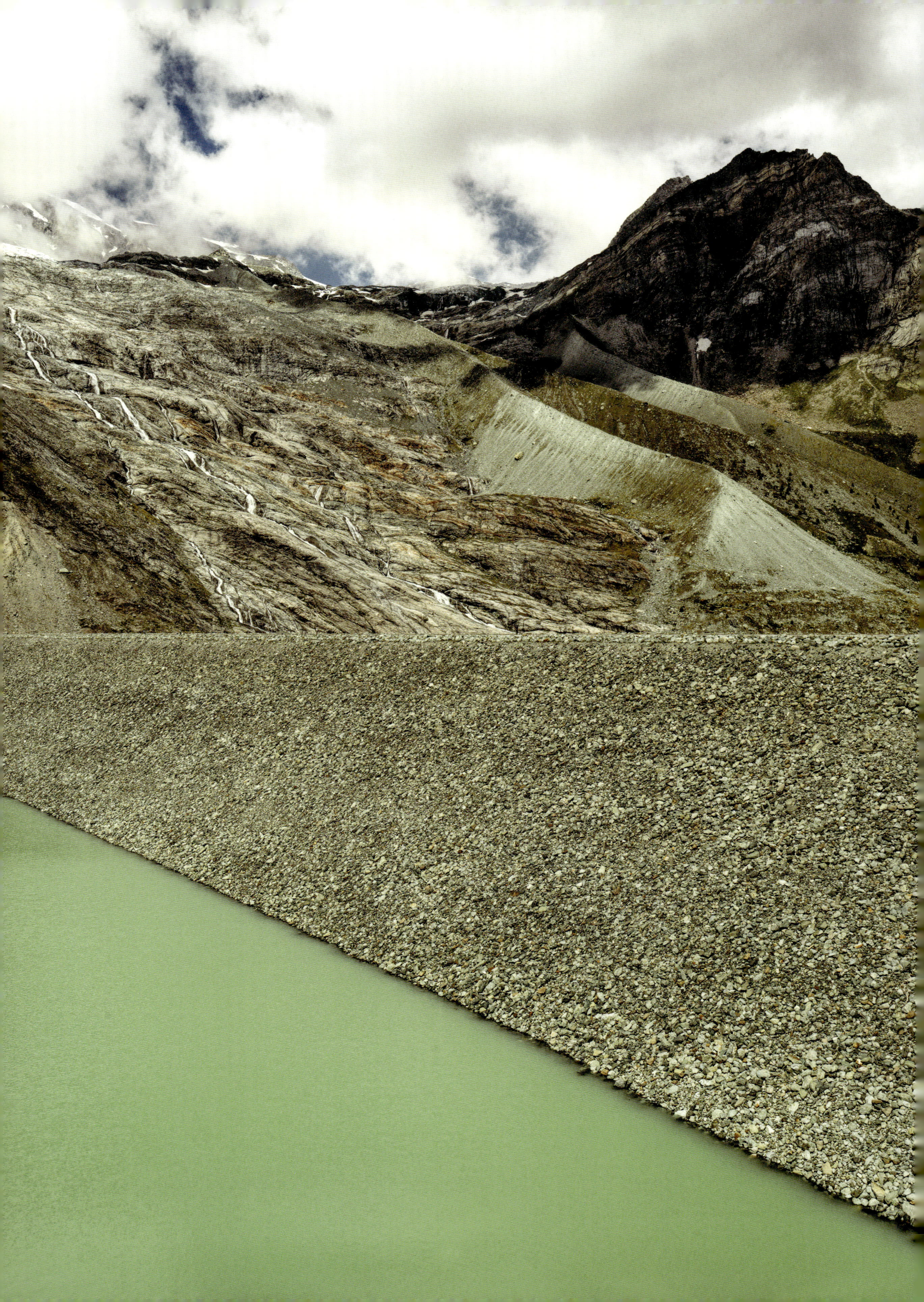

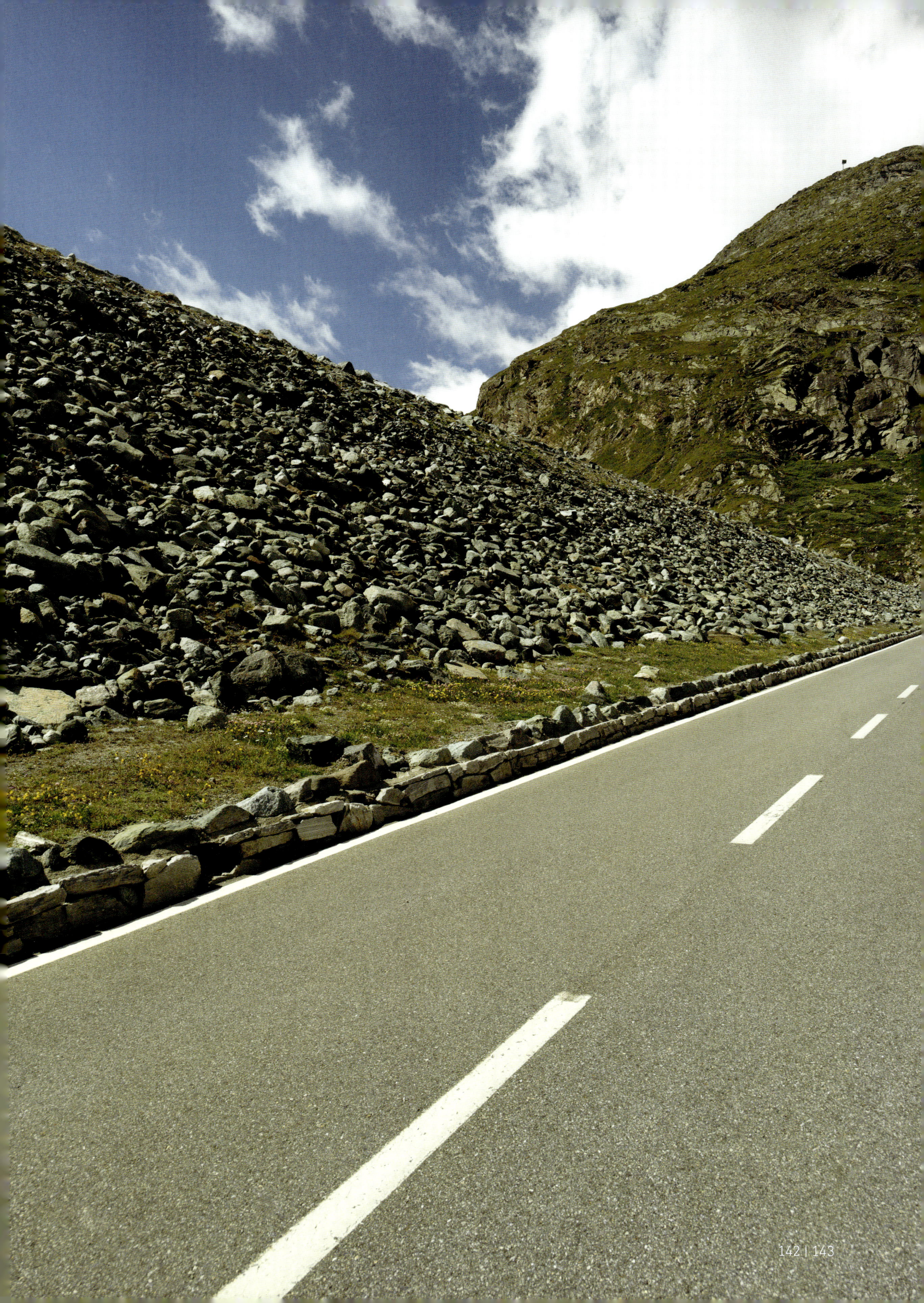

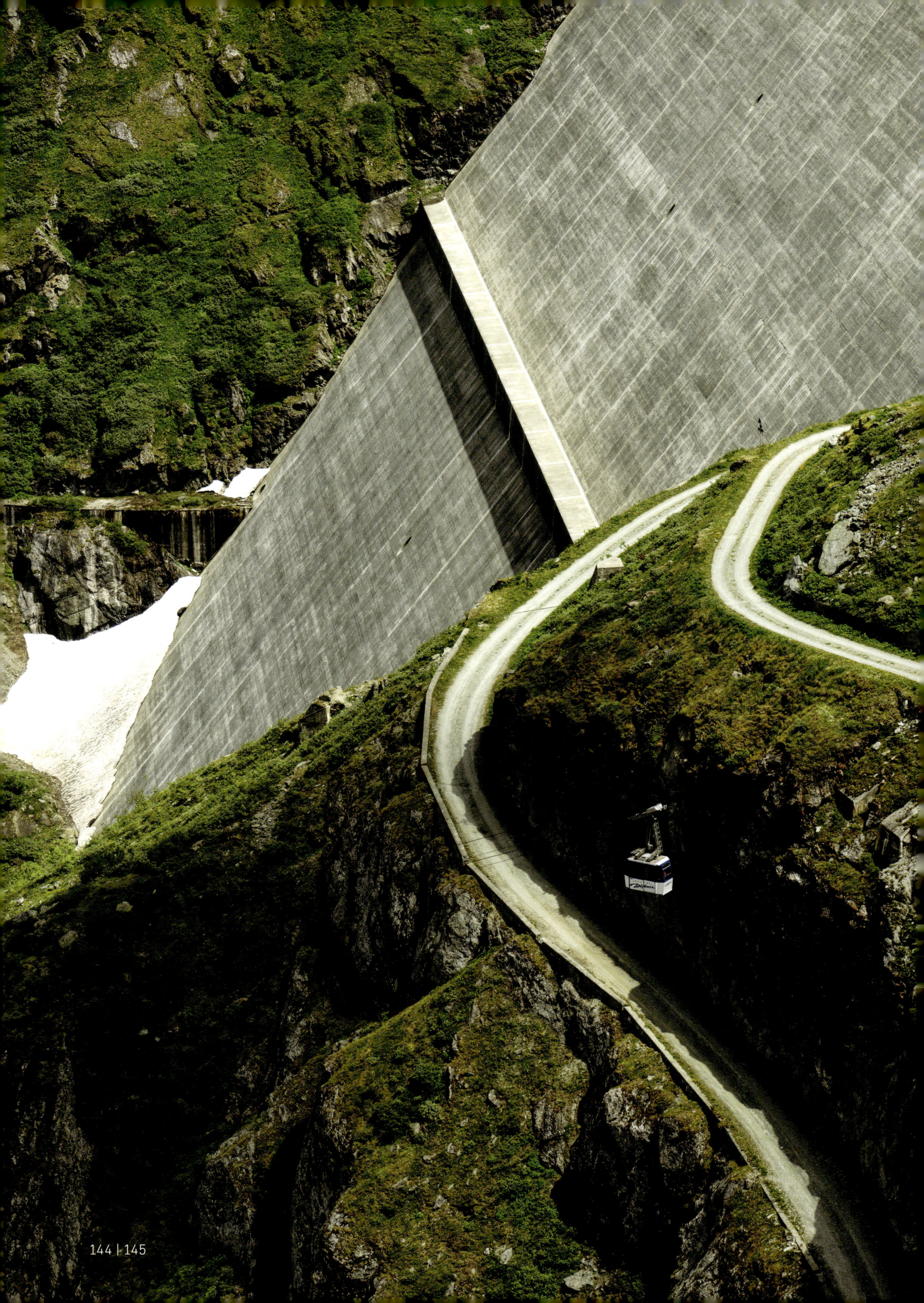

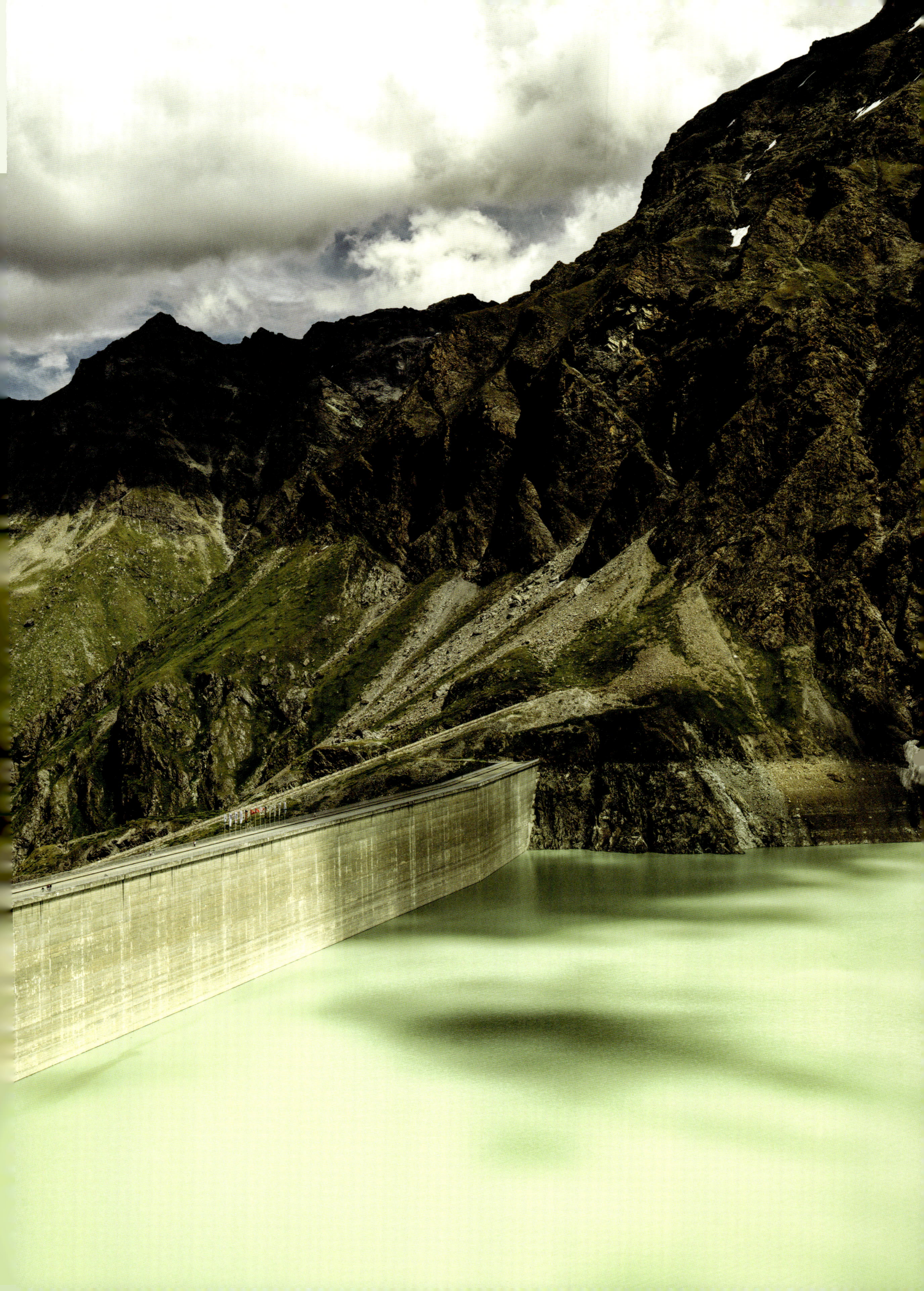

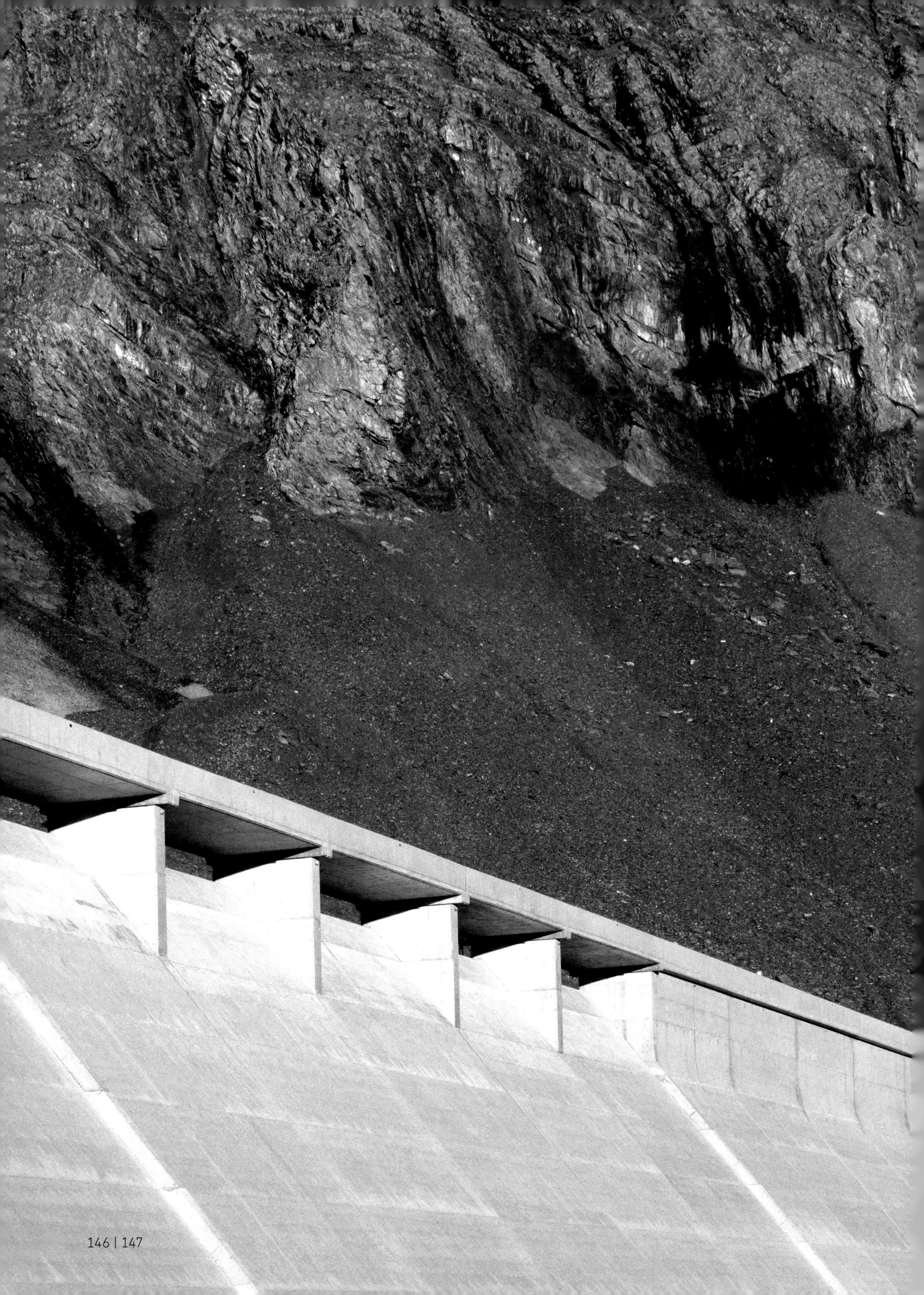

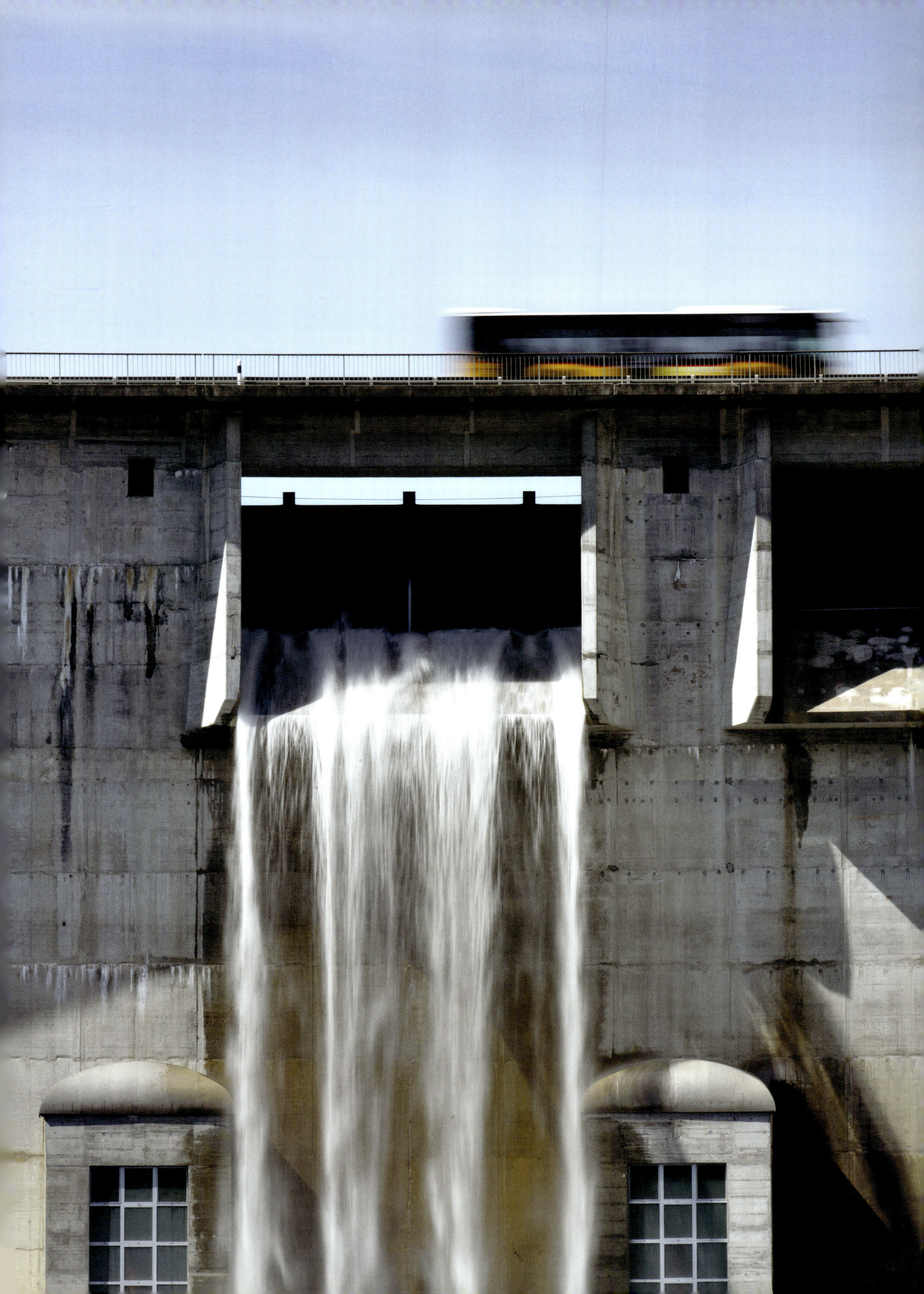

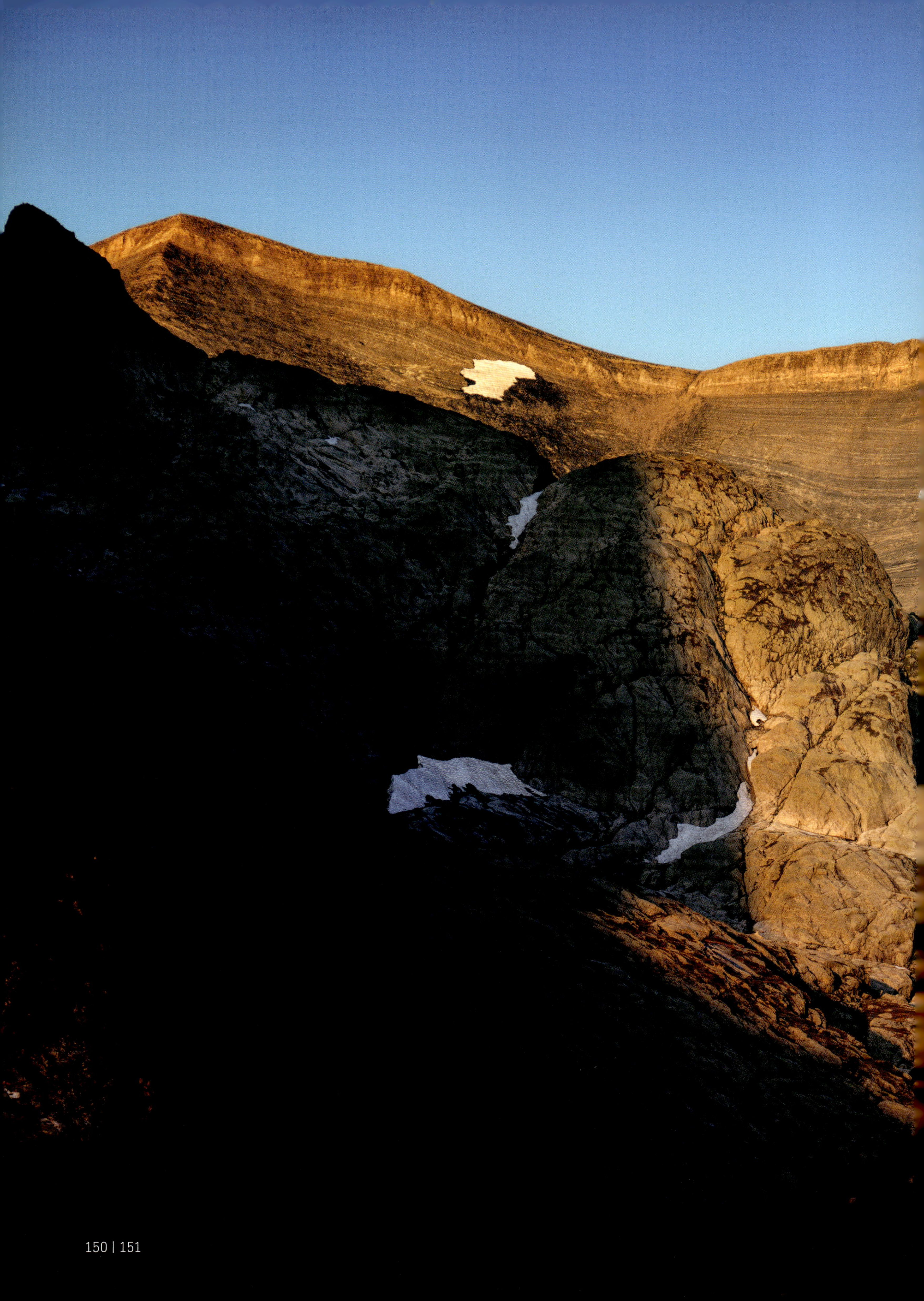

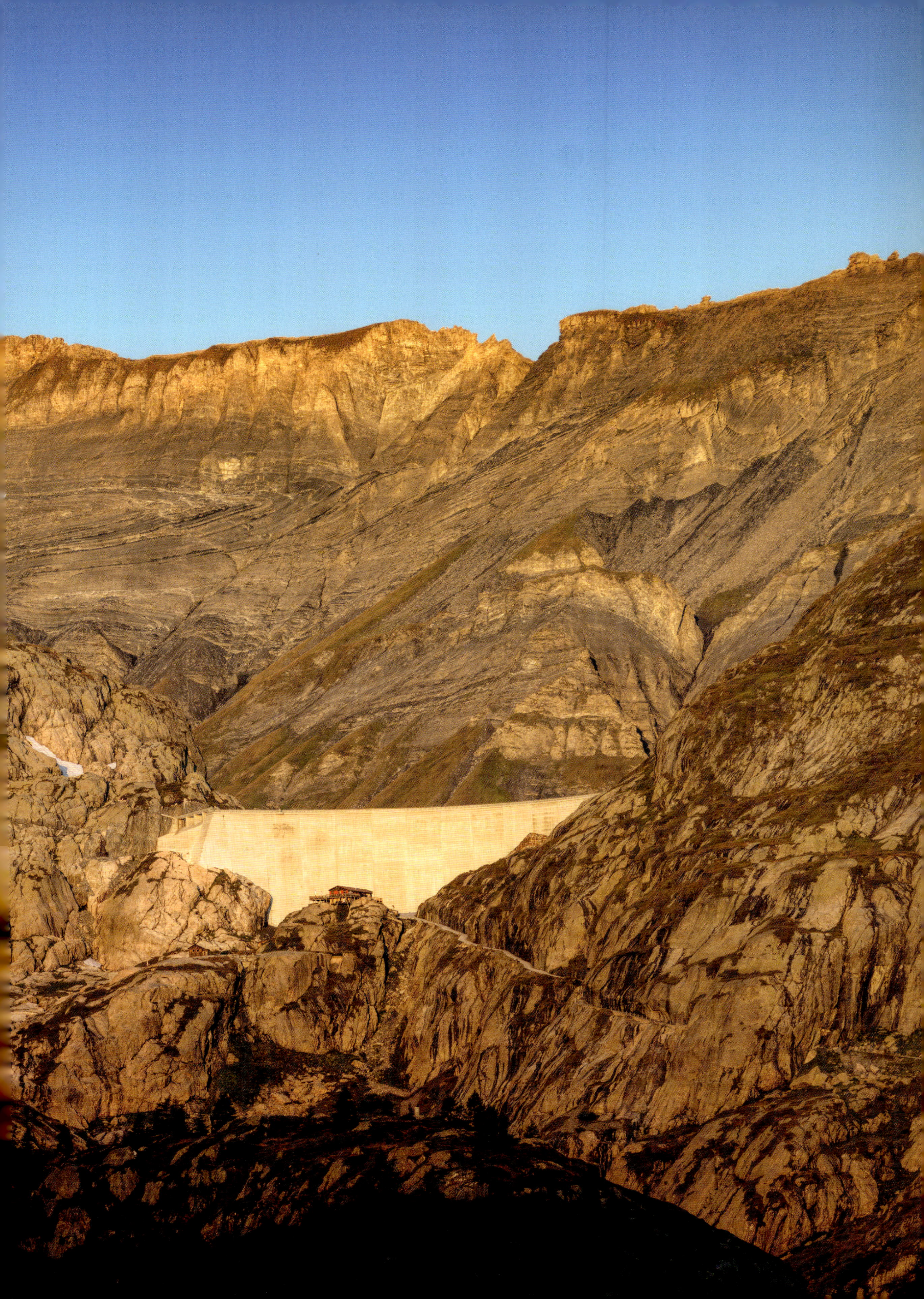

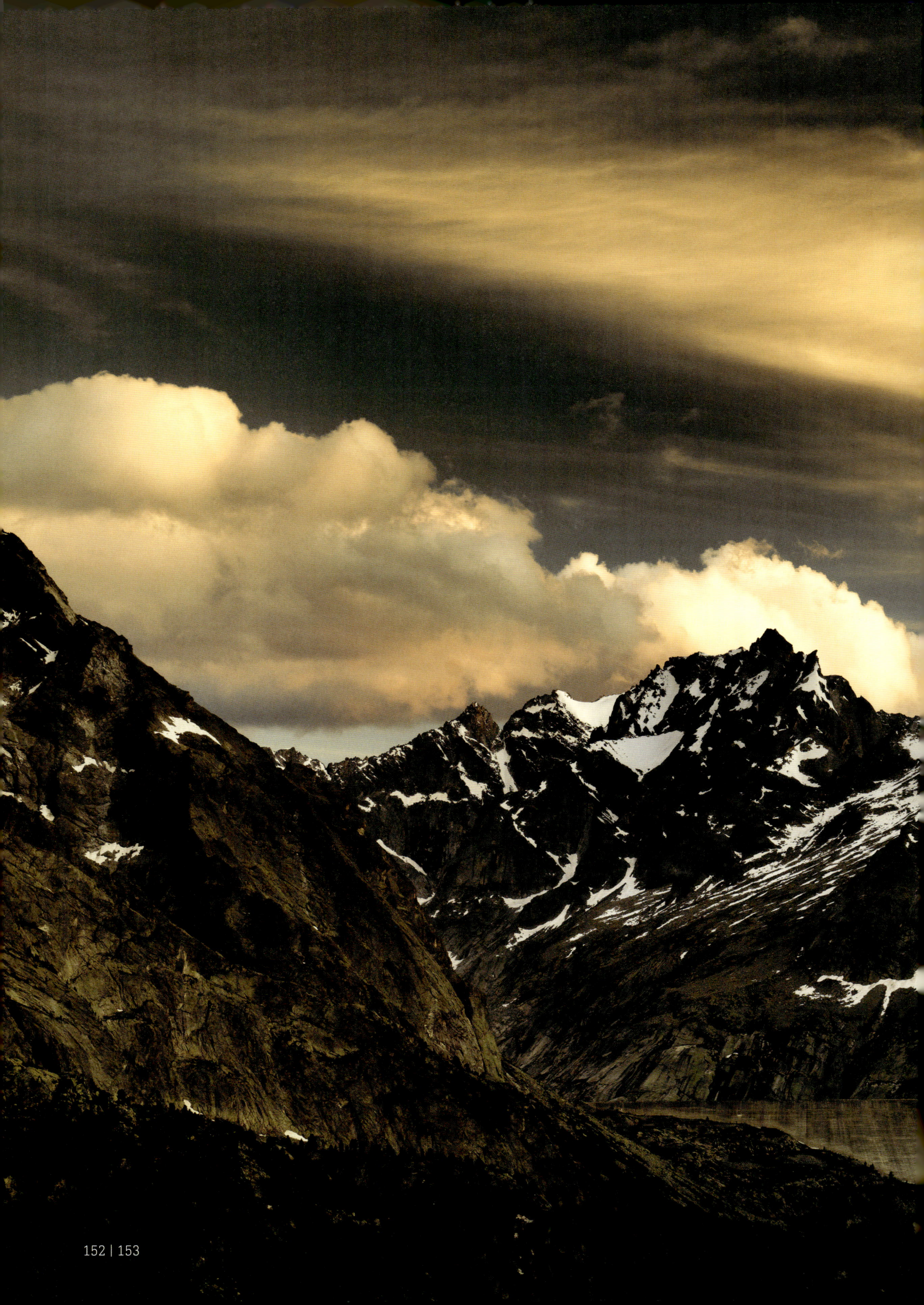

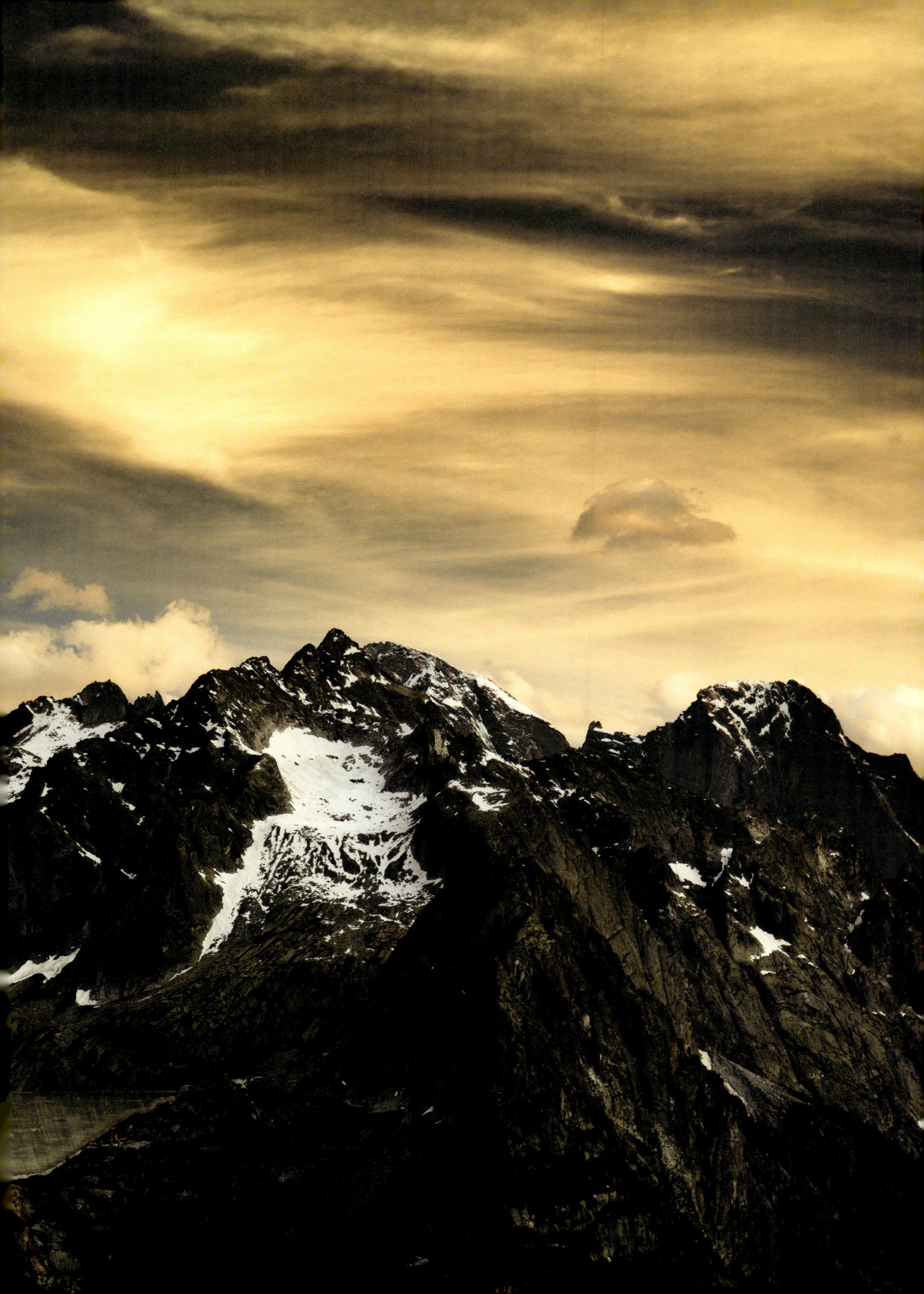

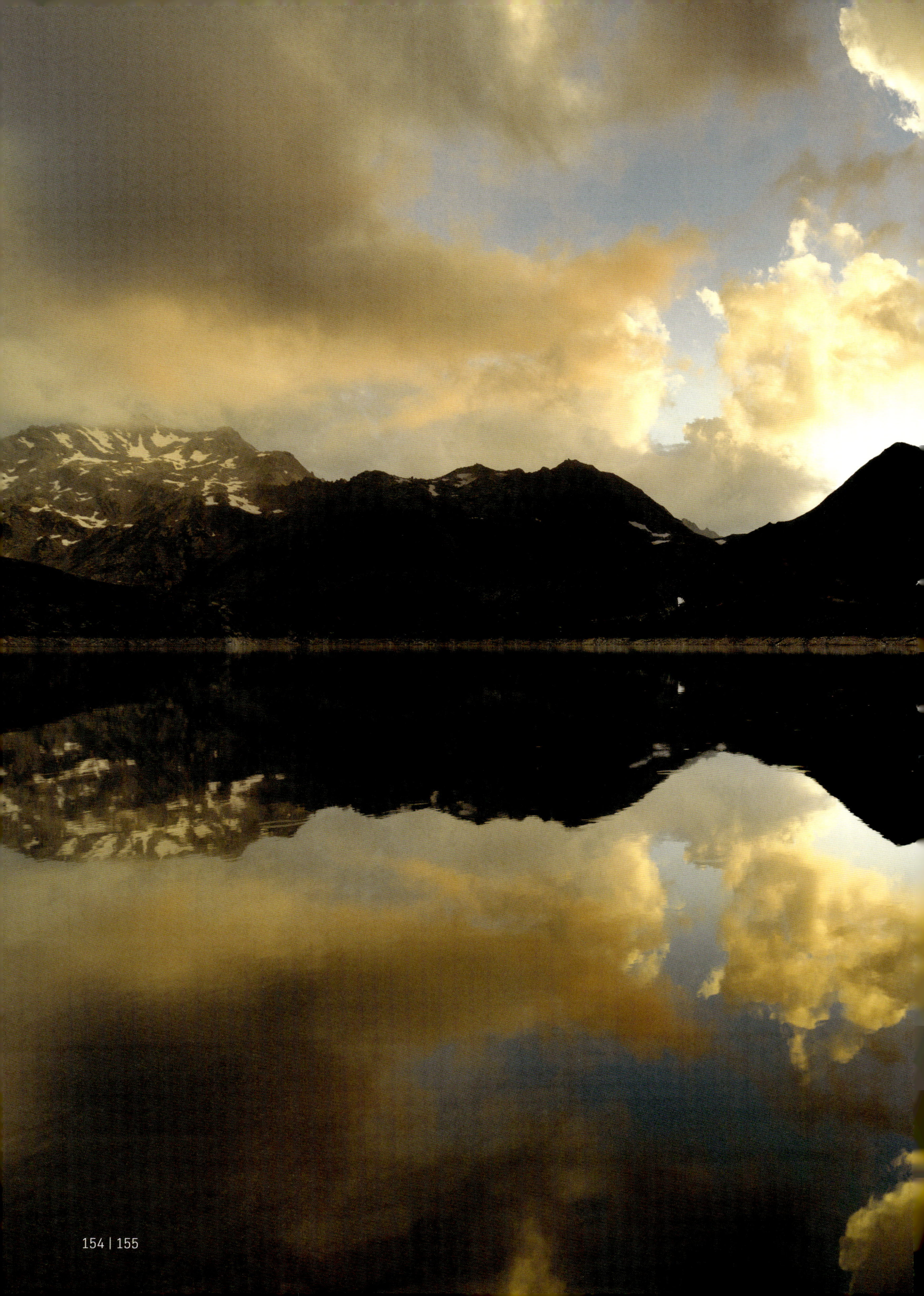

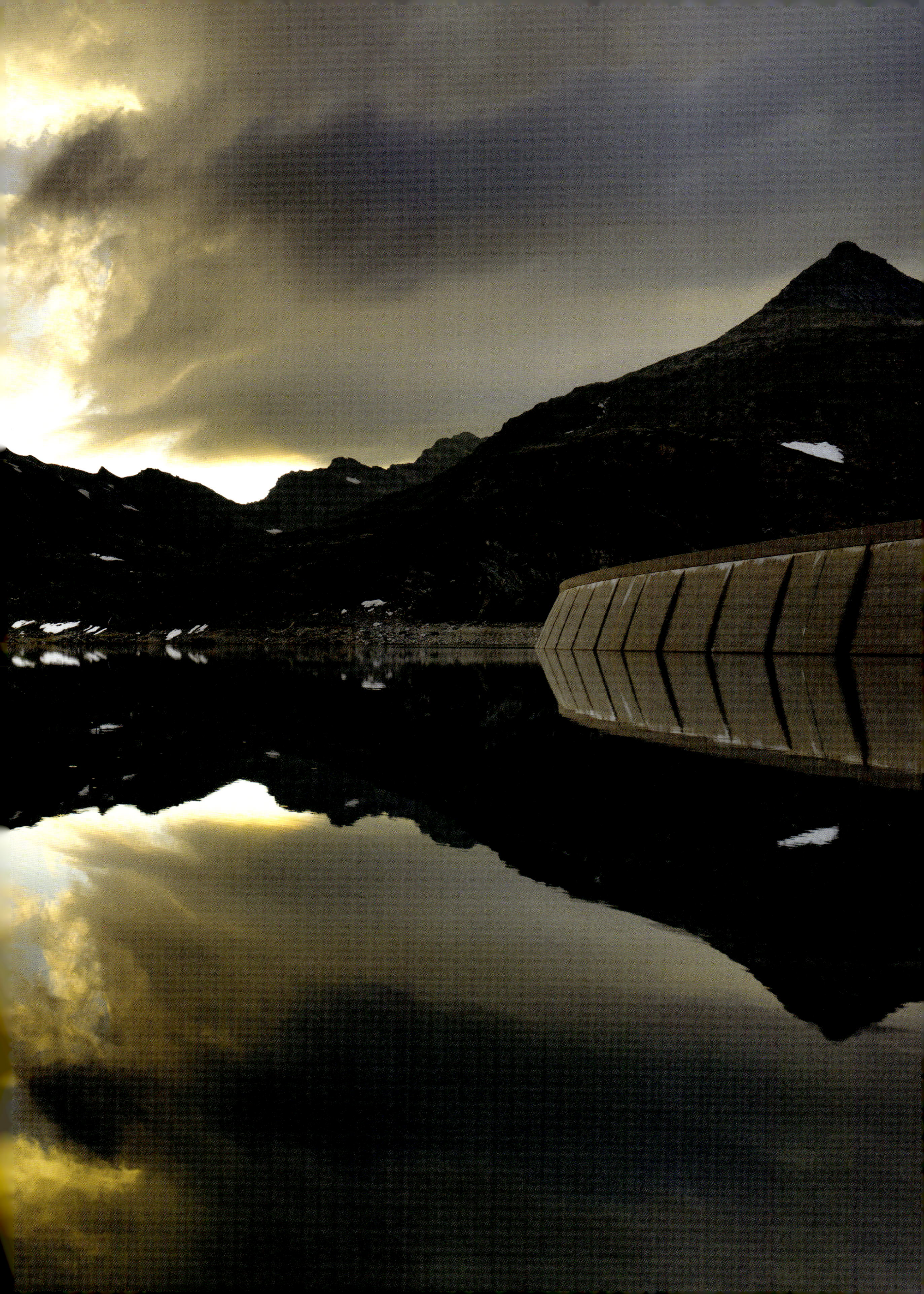

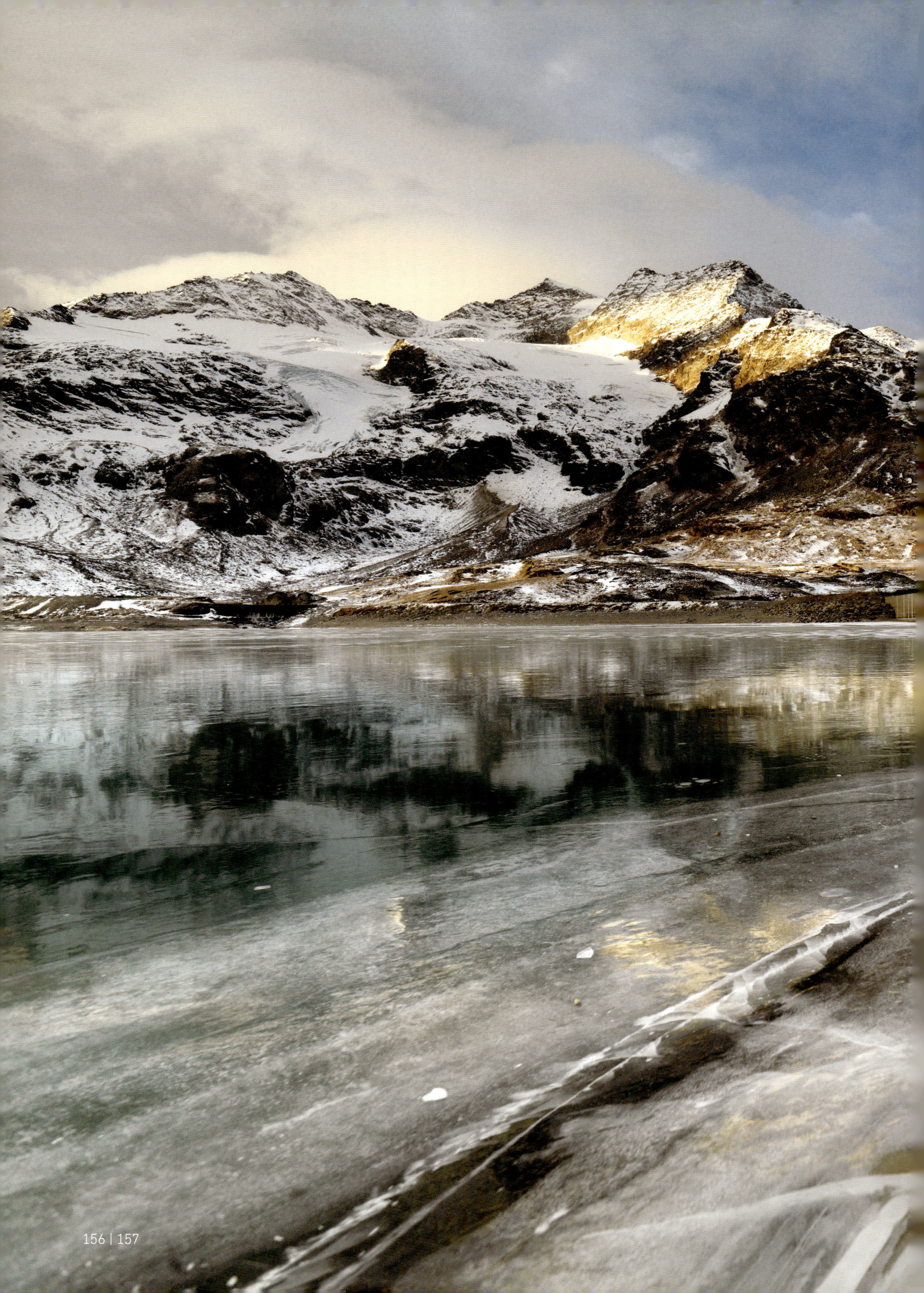

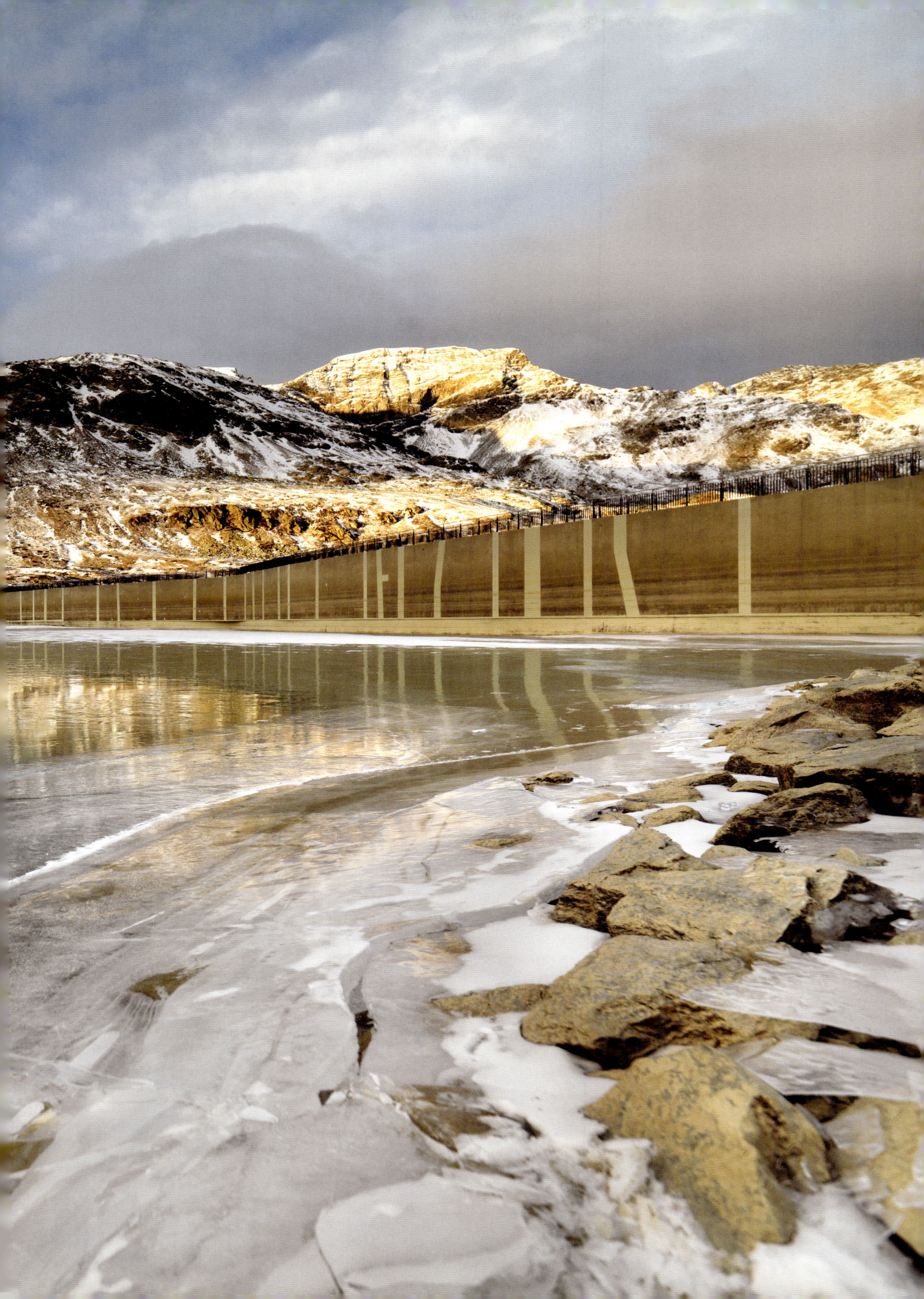

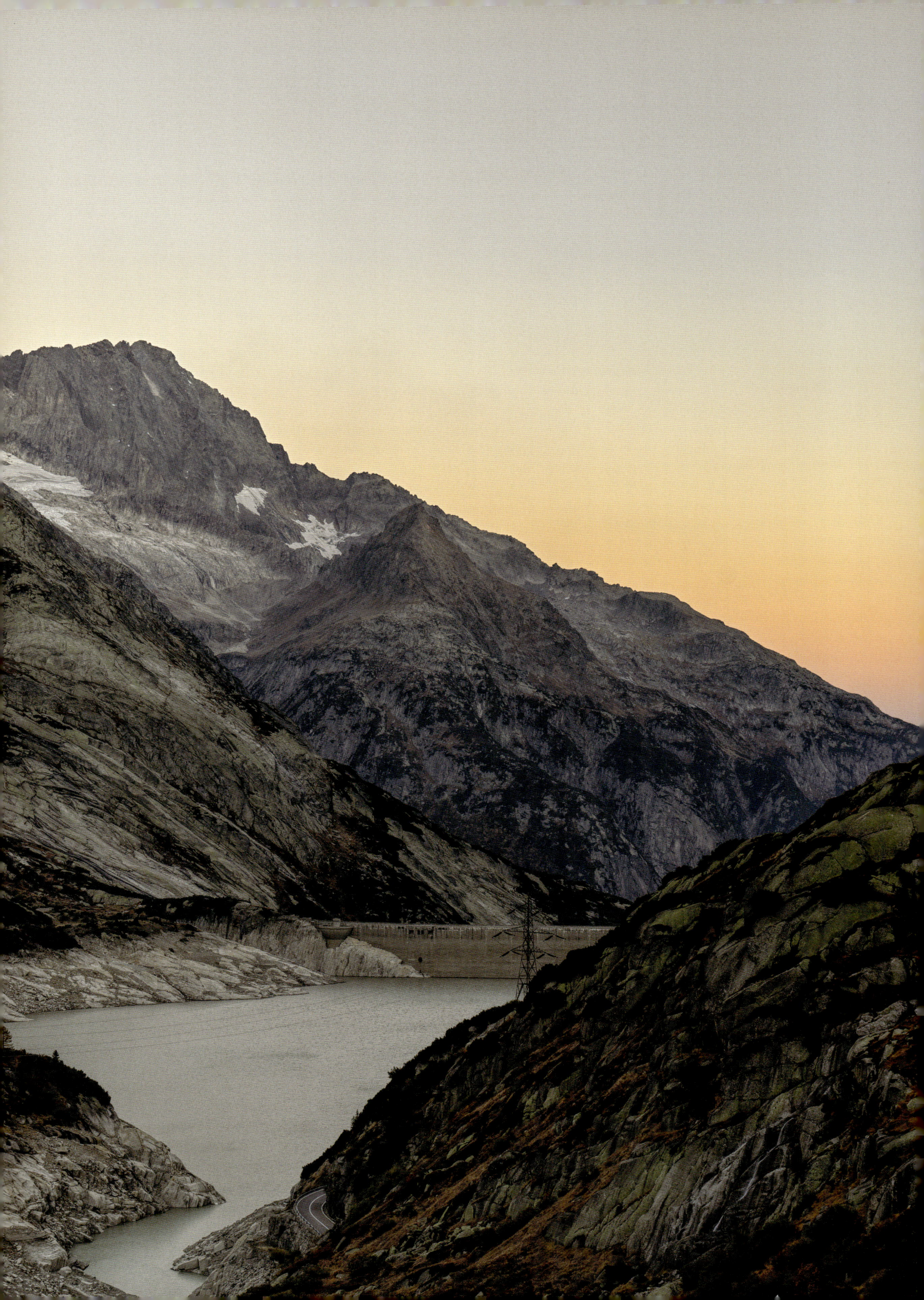

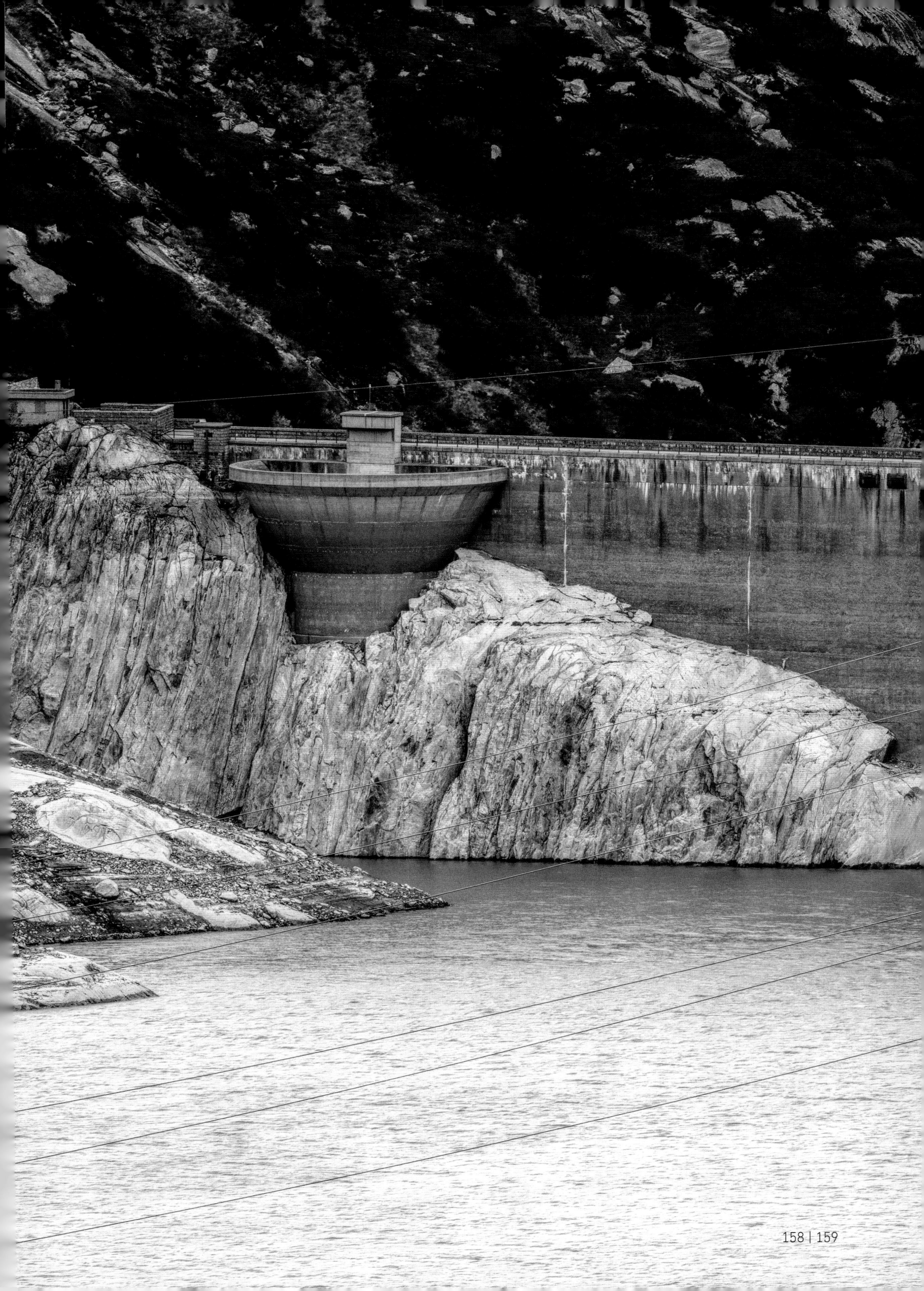

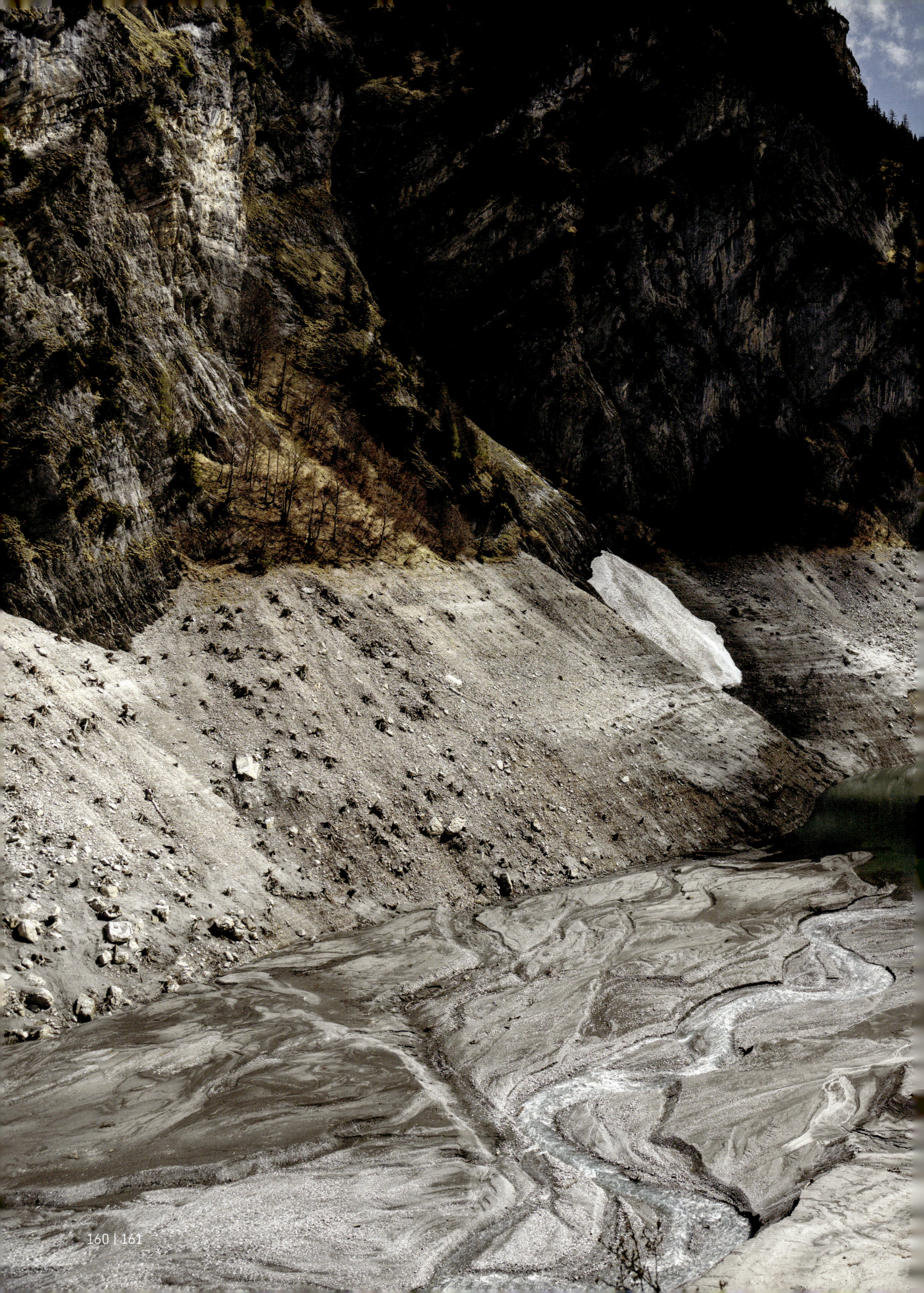

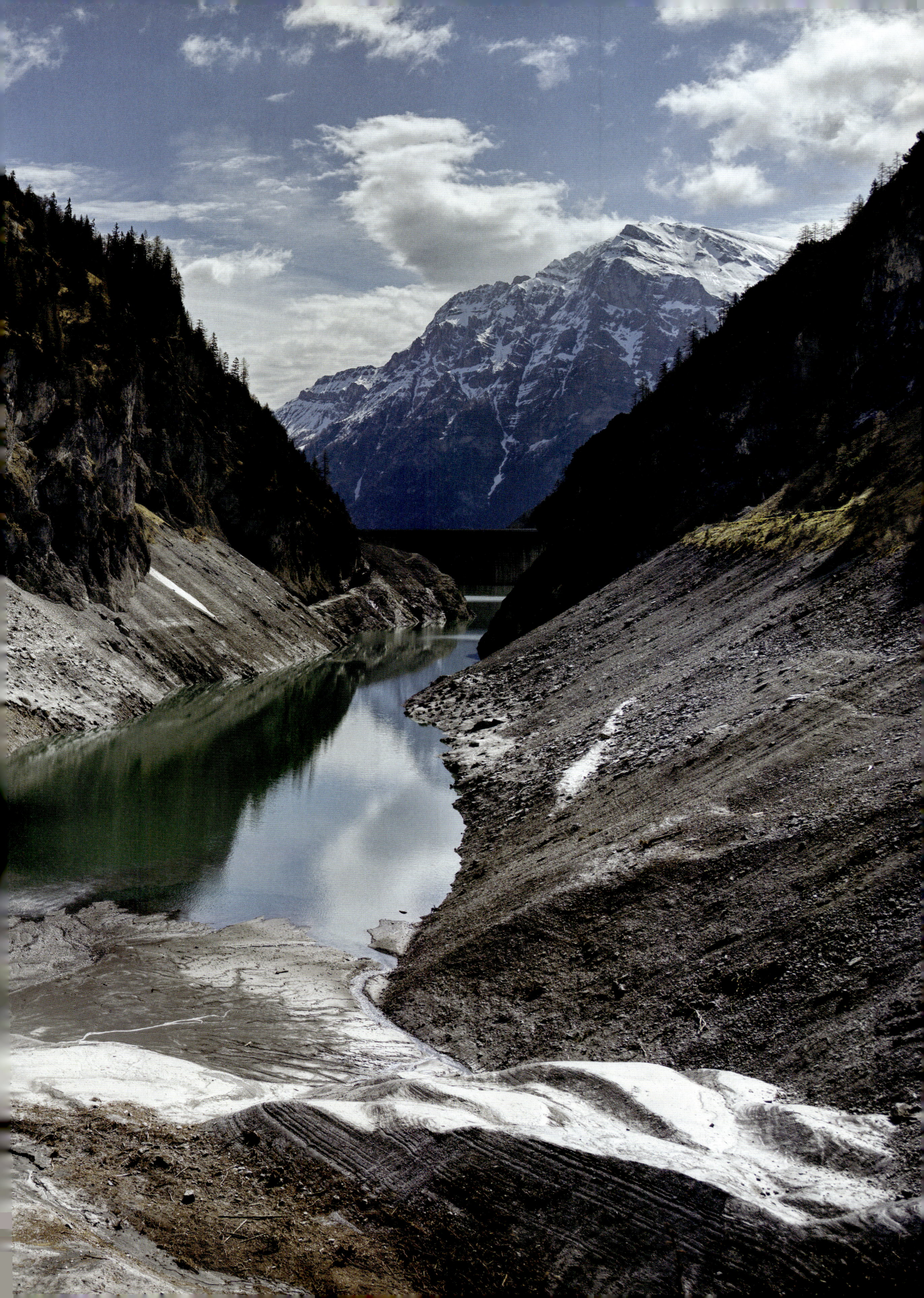

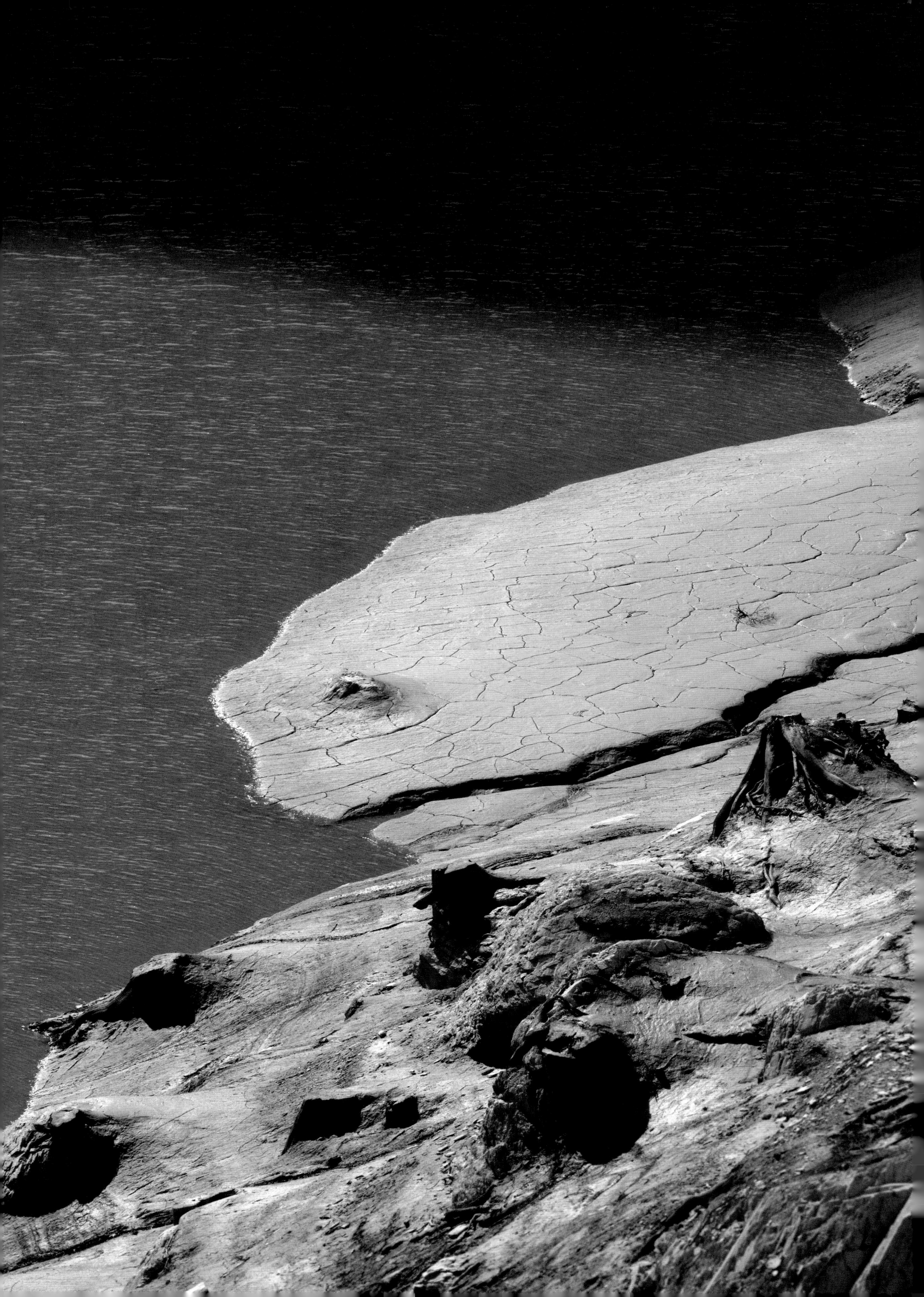

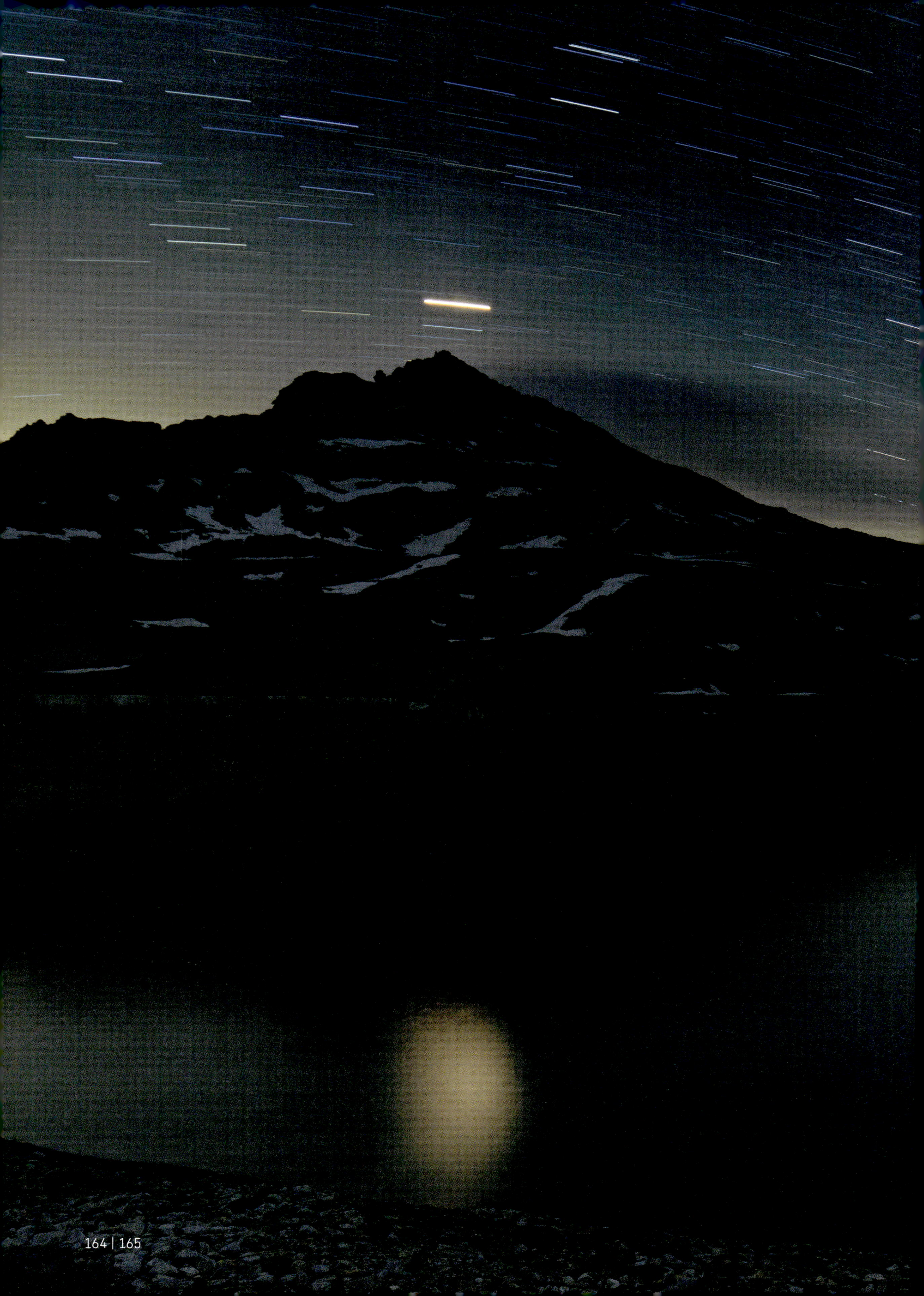

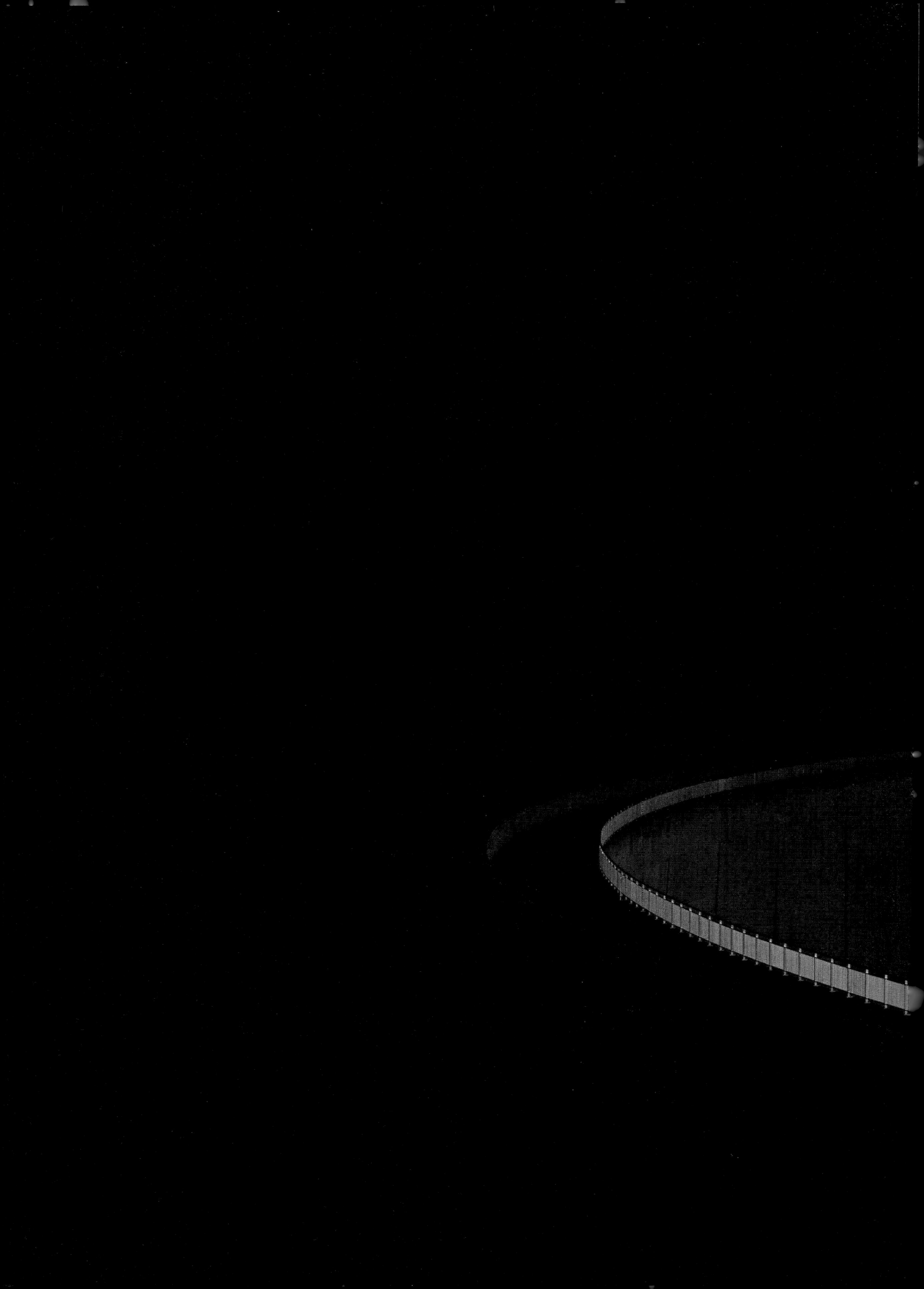

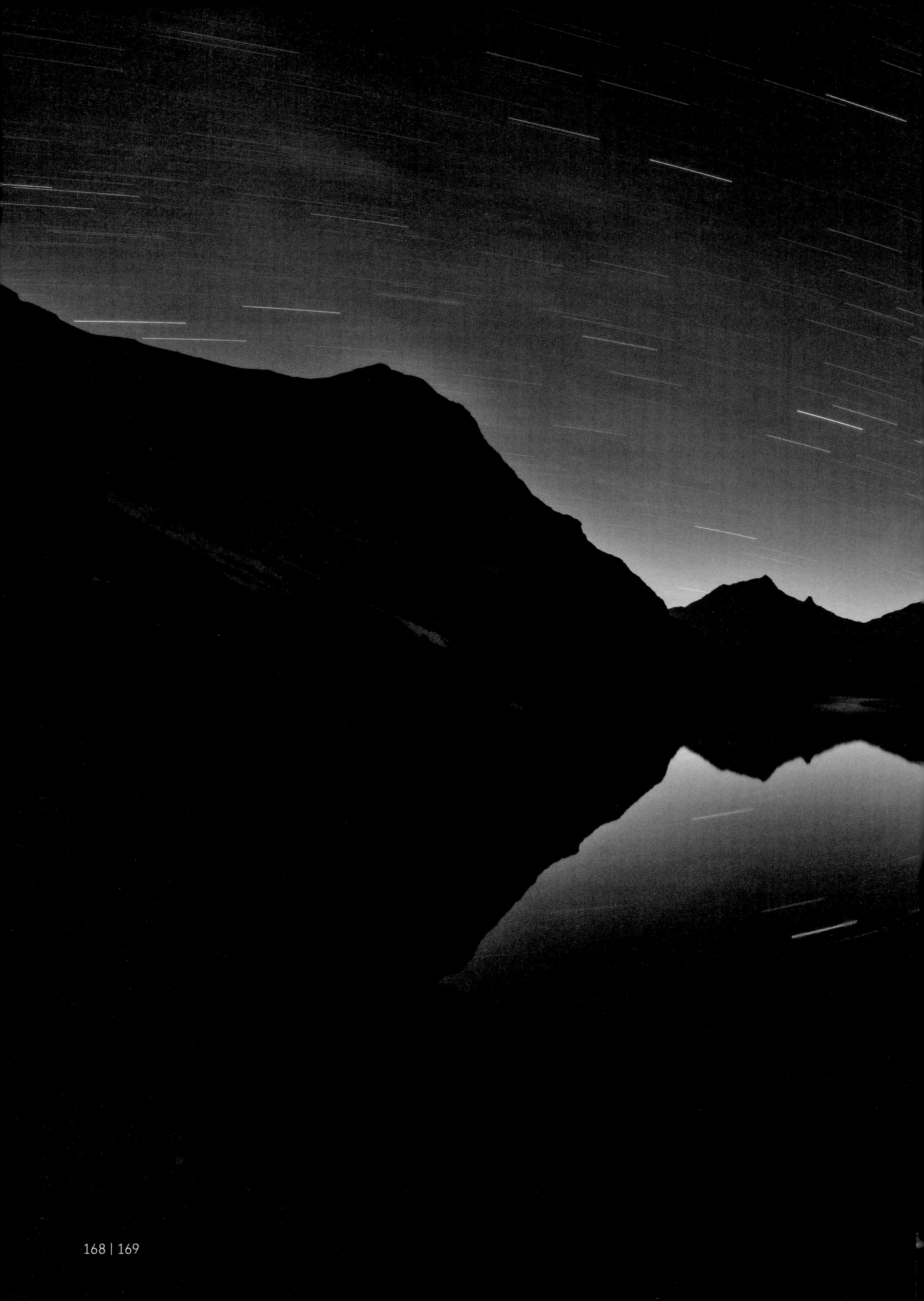

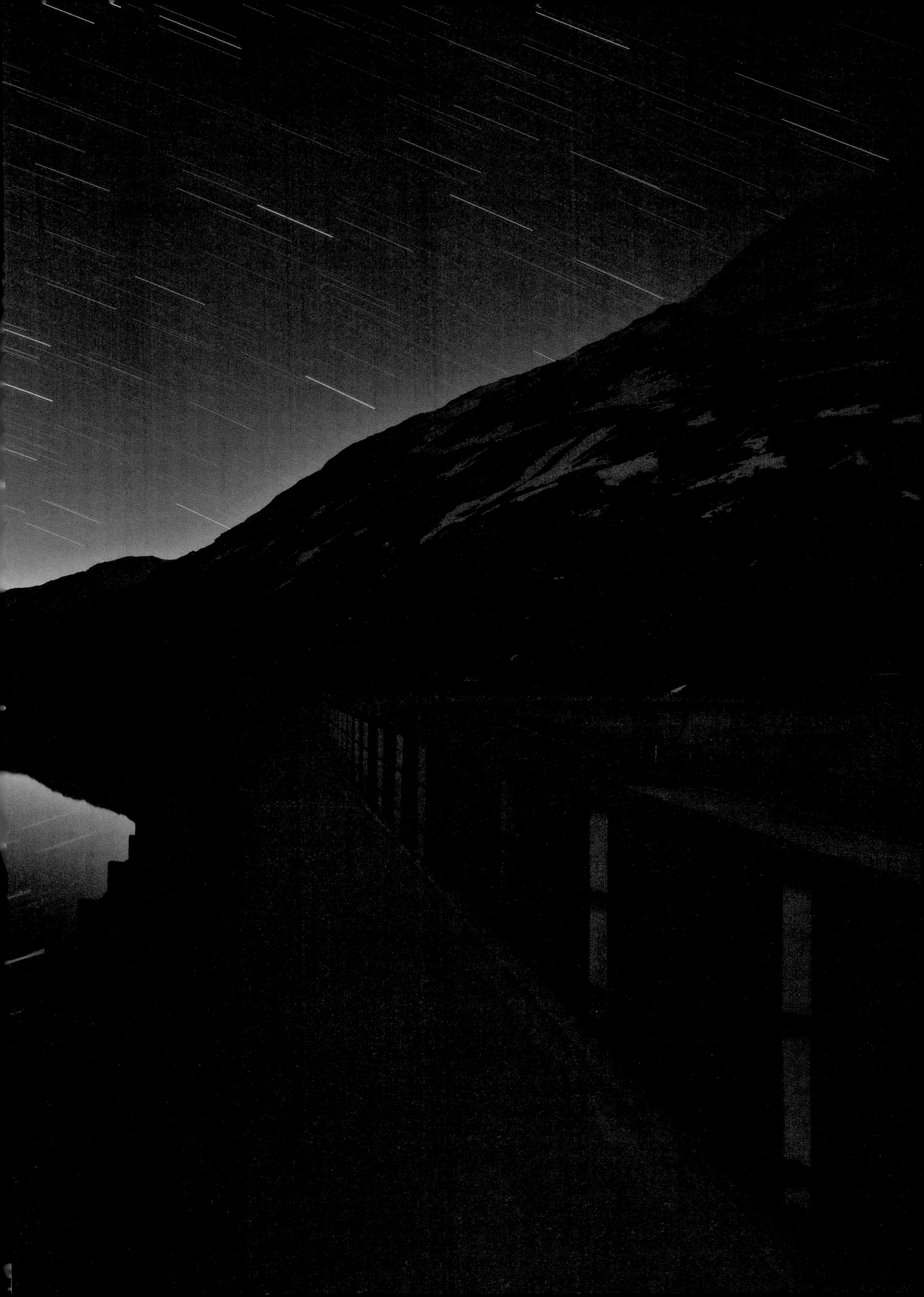

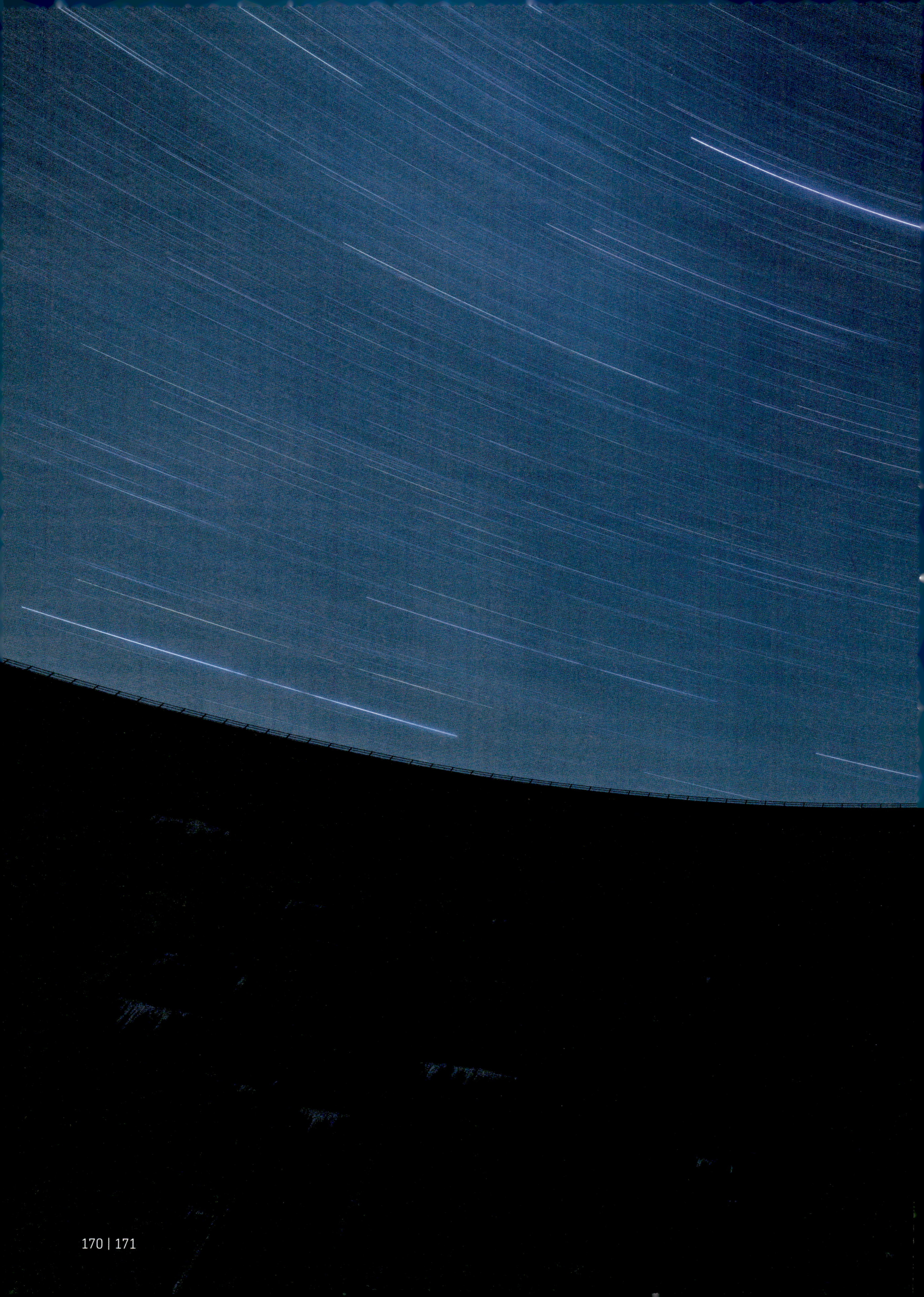

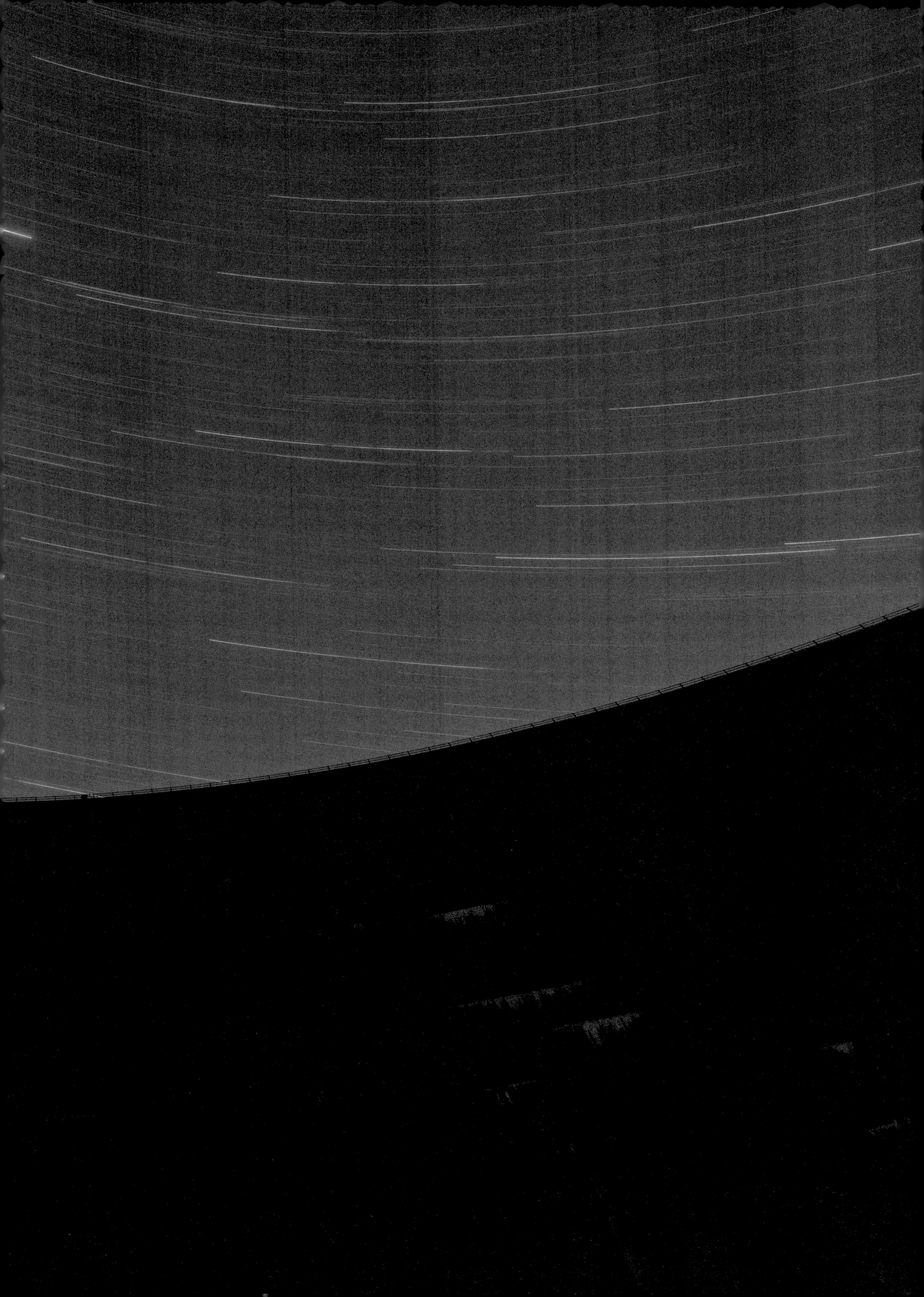

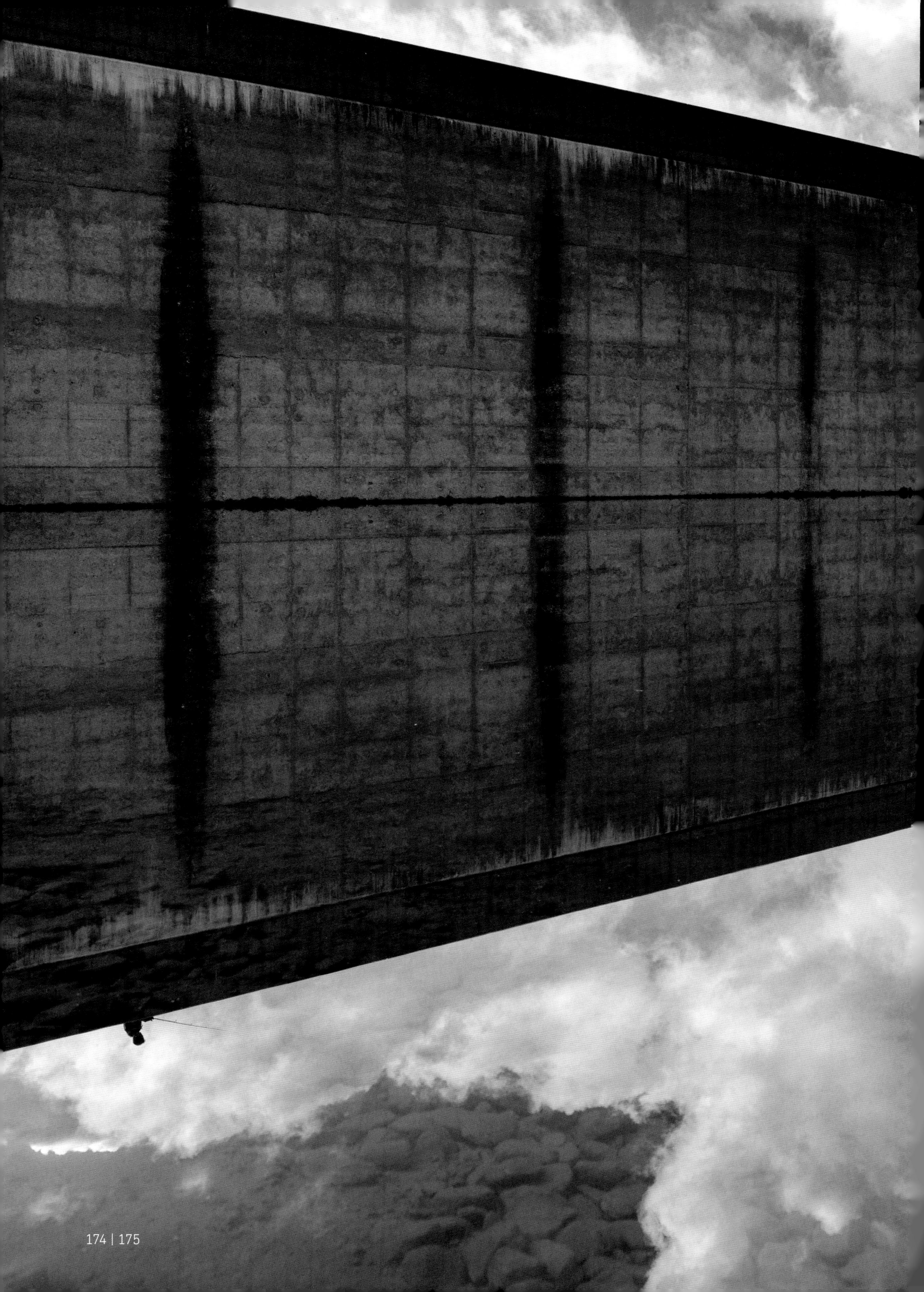

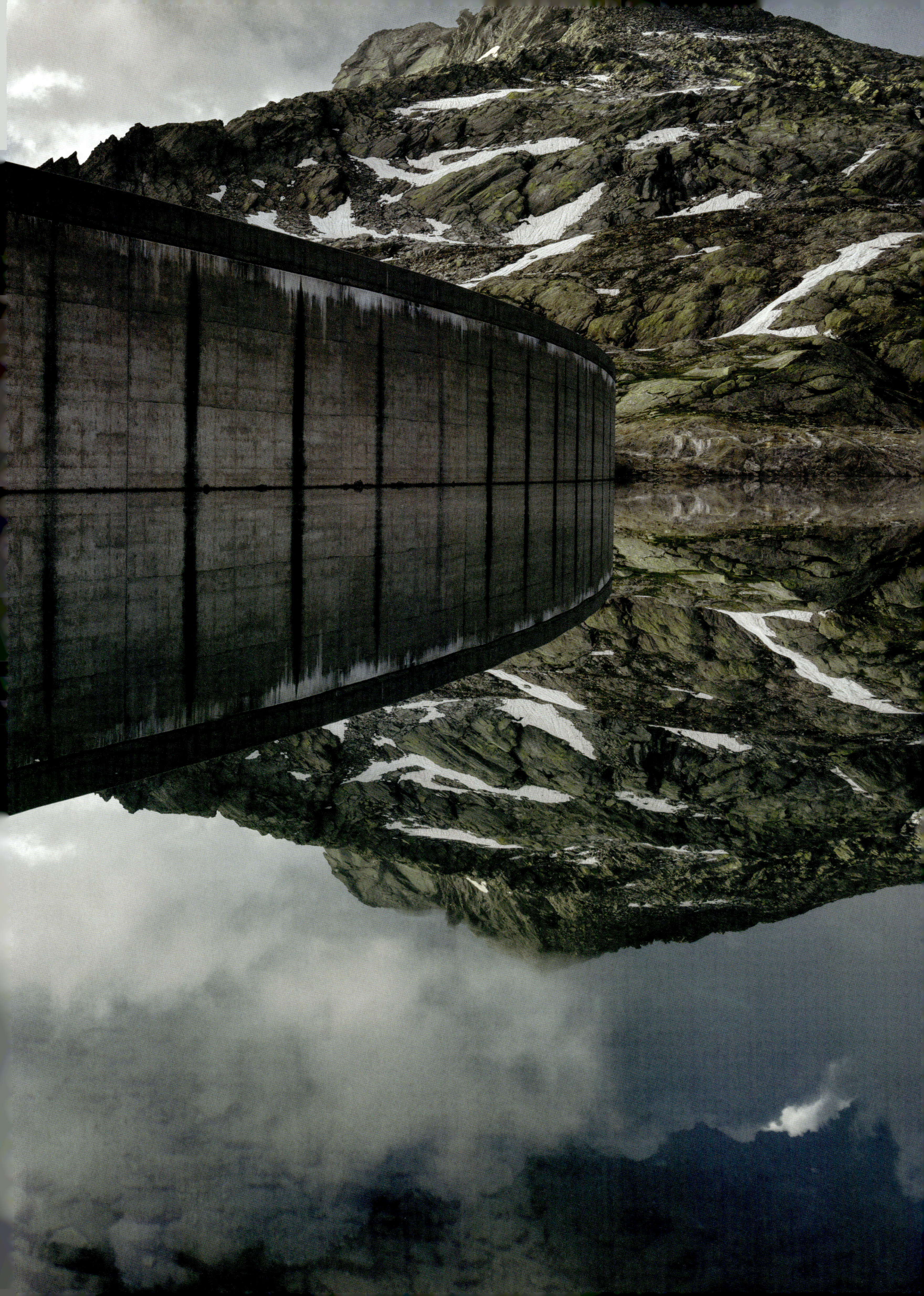

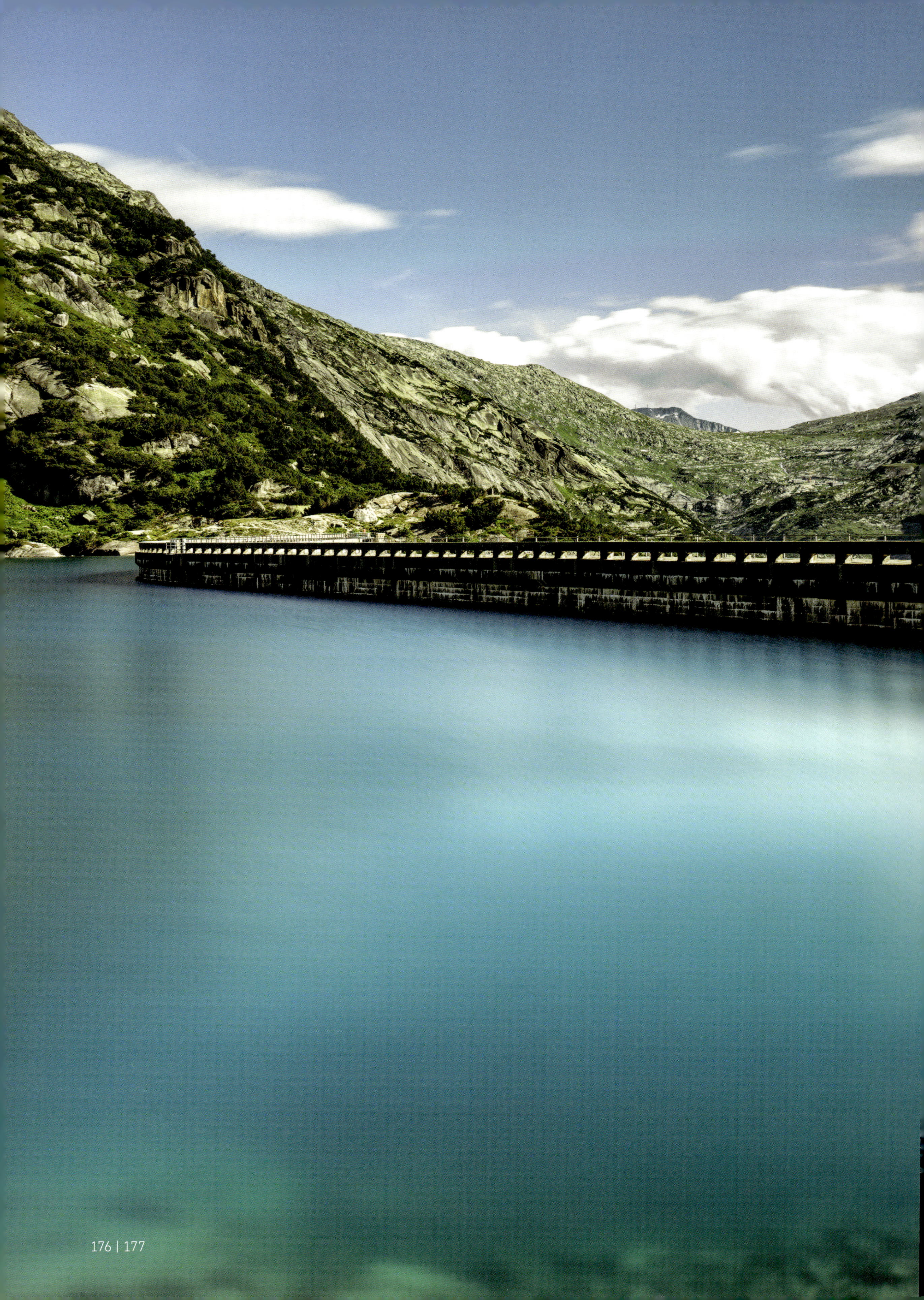

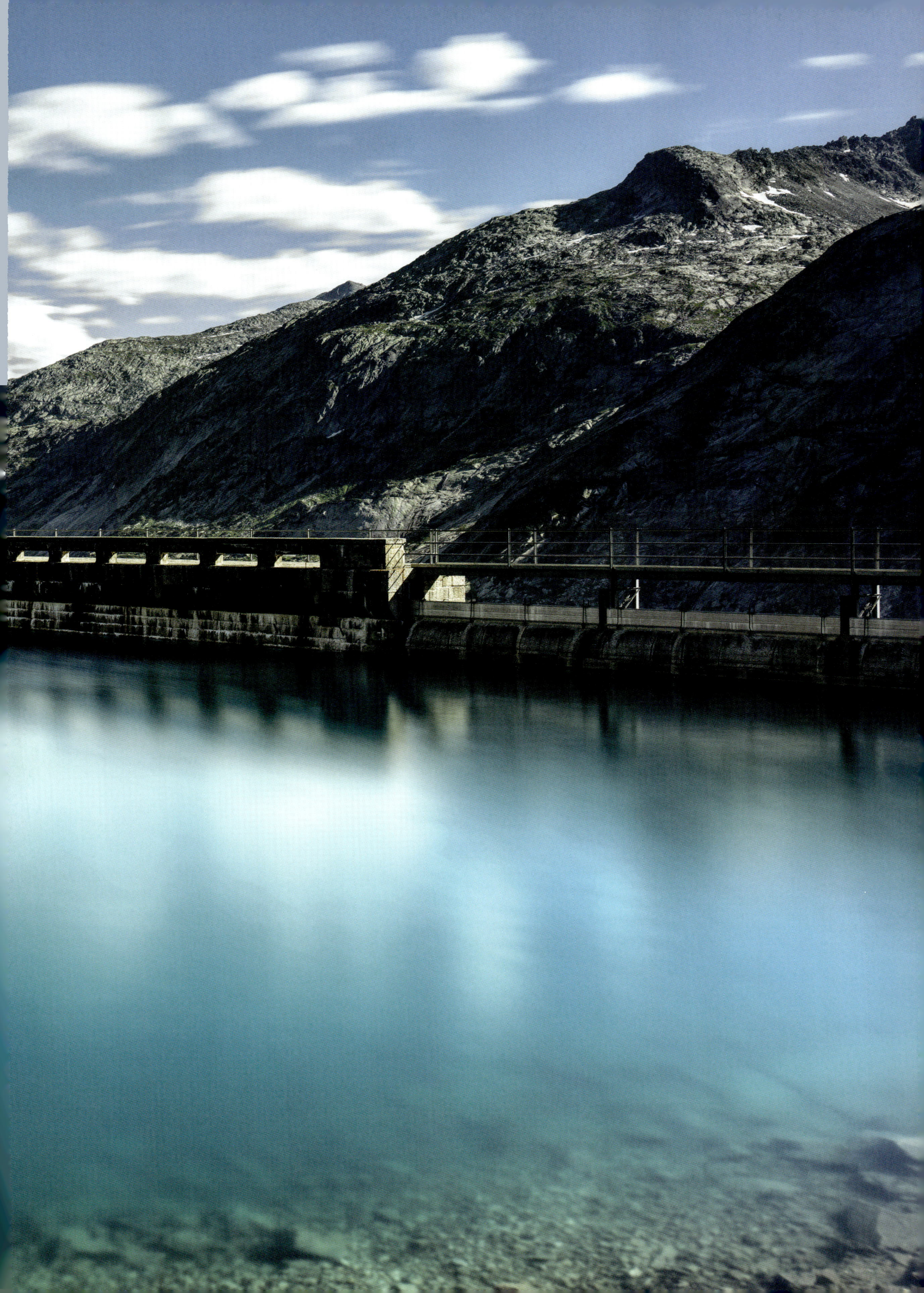

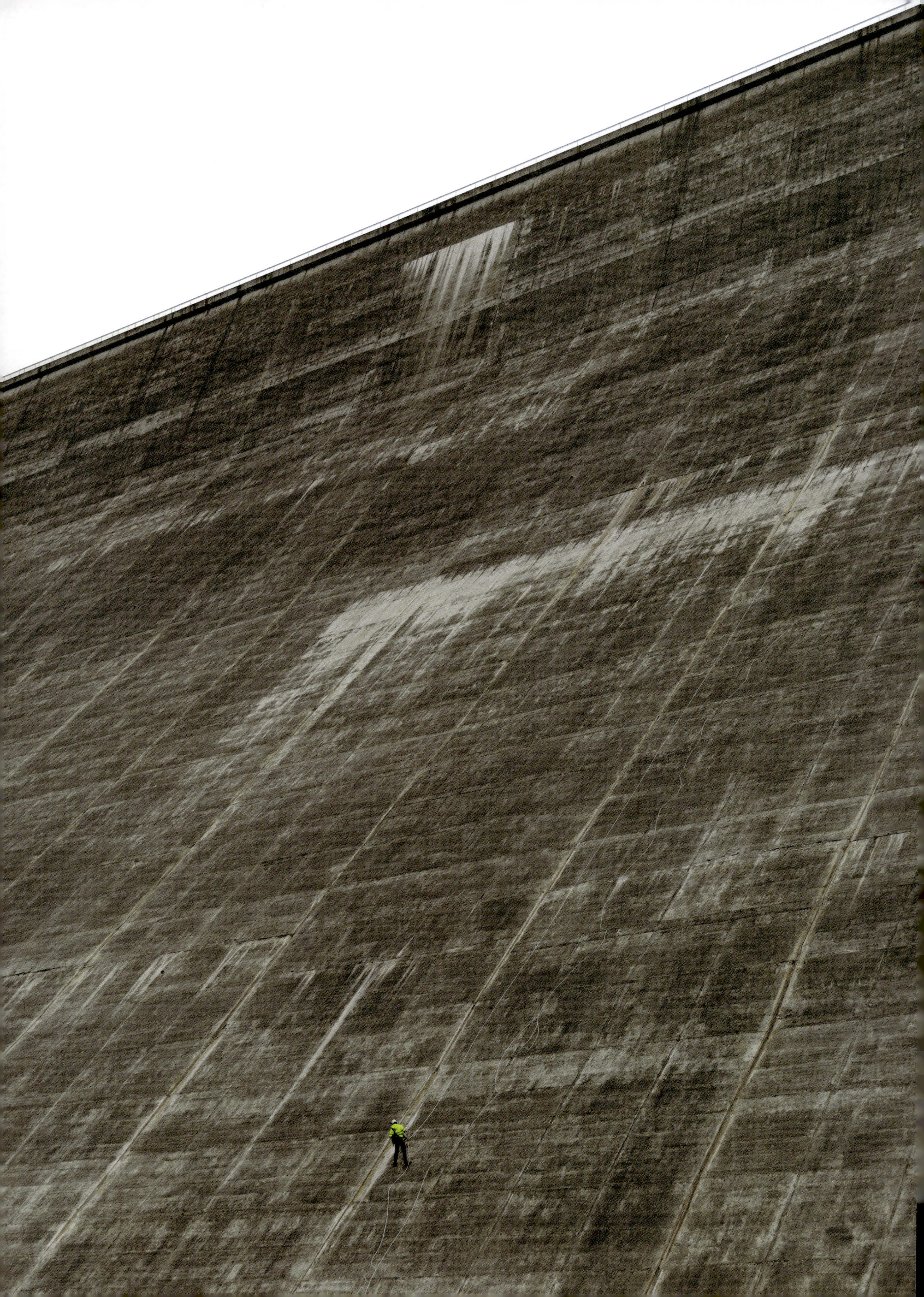

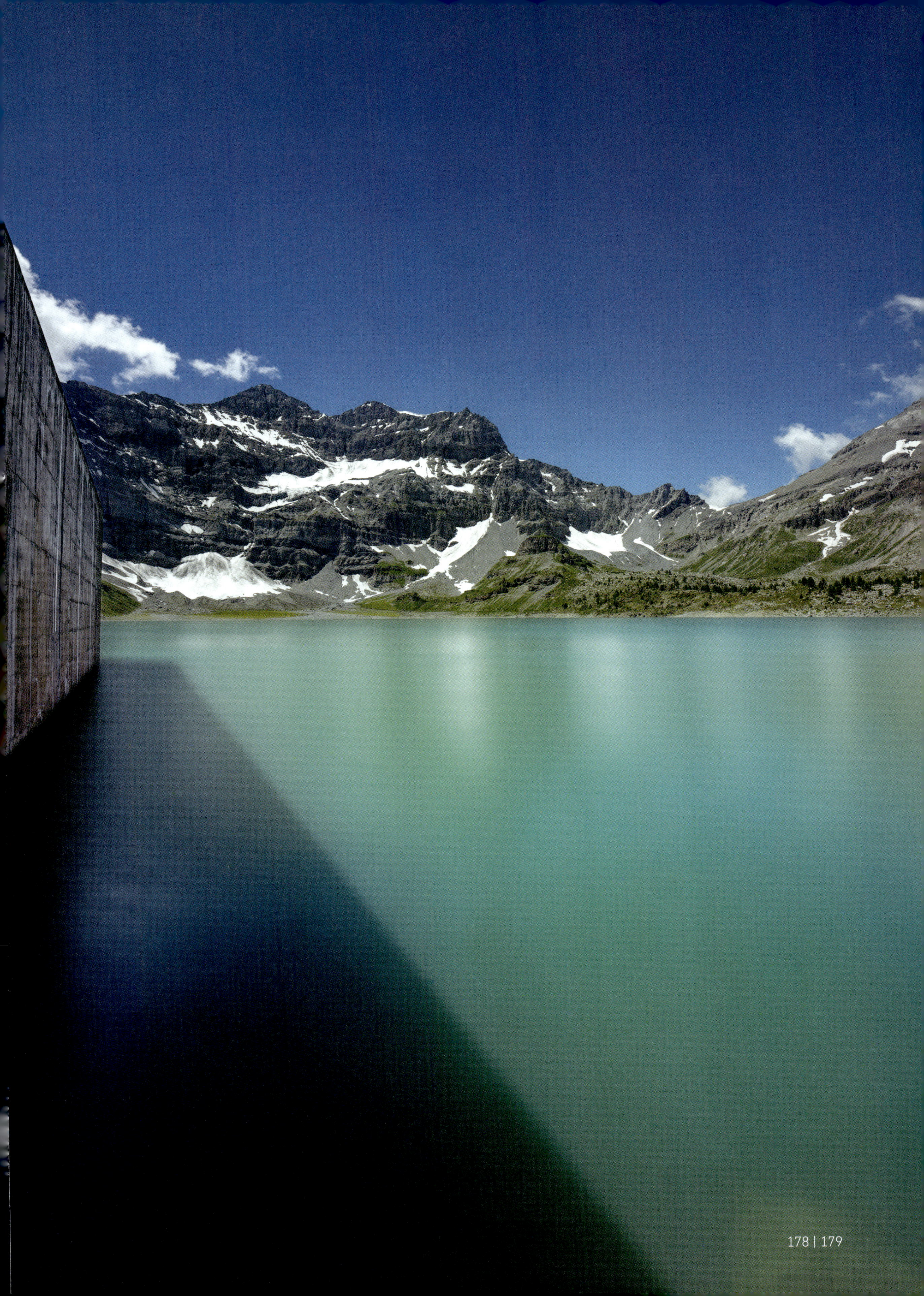

Lägh da l'Albigna

Stauanlage | dam system	
Standort | location	Bregaglia
Baujahr | construction year	1959
Sperrentyp | dam type	Schwergewichtsmauer | gravity dam
Bauvolumen | construction volume	940'000 m³
Sperrenhöhe | dam height	115 m
Kronenlänge | crest length	759 m
Kronenbreite | crest width	7 m
Fundamentbreite | foundation width	83 m
Kronenkote | crest elevation	2165 m ü. M. | MASL
Stausee | reservoir lake	
Stauvolumen | reservoir volume	69'000'000 m³
Oberfläche | surface	1'240'000 m²

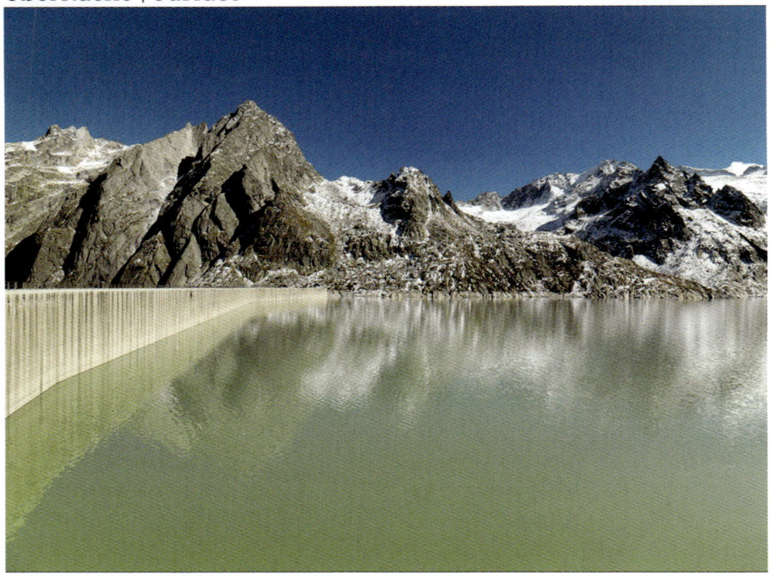

Lago Bianco (Nord)

Stauanlage | dam system	
Standort | location	Pontresina, Poschiavo
Baujahr | construction year	1911
Sperrentyp | dam type	Gewichtsmauer | gravity wall
Bauvolumen | construction volume	9'000 m³
Sperrenhöhe | dam height	13 m
Kronenlänge | crest length	290 m
Kronenbreite | crest width	2.5 m
Fundamentbreite | foundation width	7 m
Kronenkote | crest elevation	2235 m ü. M. | MASL
Stausee | reservoir lake	
Stauvolumen | reservoir volume	18'600'000 m³
Oberfläche | surface	1'500'000 m²

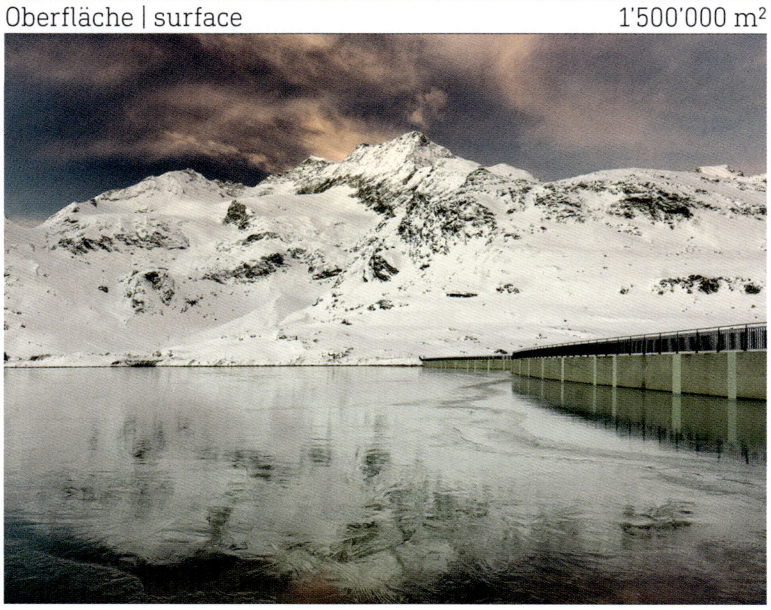

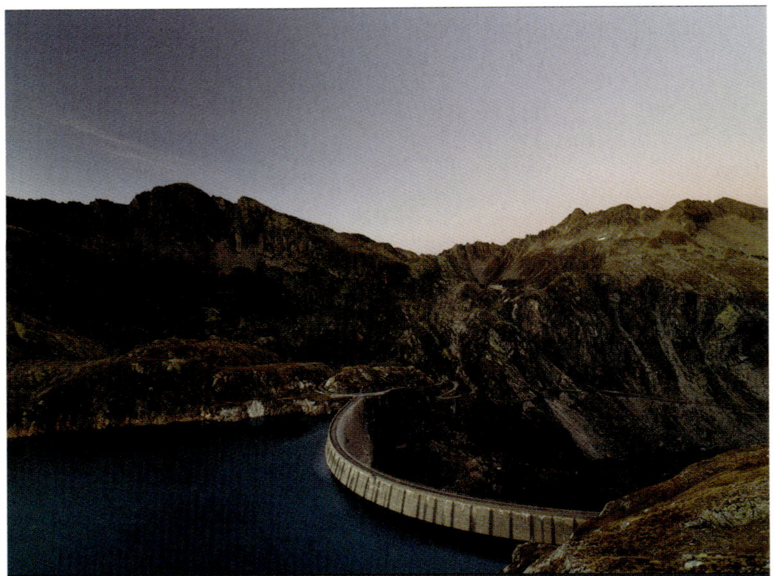

Lago dei Cavagnöö

Stauanlage | dam system	
Standort | location	Cevio
Baujahr | construction year	1968
Sperrentyp | dam type	Bogenmauer | arch dam
Bauvolumen | construction volume	223'000 m³
Sperrenhöhe | dam height	111 m
Kronenlänge | crest length	317 m
Kronenbreite | crest width	5 m
Fundamentbreite | foundation width	15.8 m
Kronenkote | crest elevation	2311 m ü. M. | MASL
Stausee | reservoir lake	
Stauvolumen | reservoir volume	29'000'000 m³
Oberfläche | surface	460'000 m²

Lac de Cleuson

Stauanlage | dam system	
Standort | location	Nendaz
Baujahr | construction year	1950
Sperrentyp | dam type	Gewichtsmauer | gravity wall
Bauvolumen | construction volume	405'000 m³
Sperrenhöhe | dam height	87 m
Kronenlänge | crest length	420 m
Kronenbreite | crest width	5.4 m
Fundamentbreite | foundation width	84.6 m
Kronenkote | crest elevation	2187 m ü. M. | MASL
Stausee | reservoir lake	
Stauvolumen | reservoir volume	19'500'000 m³
Oberfläche | surface	500'000 m²

Lai da Curnera

Stauanlage | dam system	
Standort | location	Tujetsch
Baujahr | construction year	1966
Sperrentyp | dam type	Bogenmauer | arch dam
Bauvolumen | construction volume	562'000 m³
Sperrenhöhe | dam height	153 m
Kronenlänge | crest length	350 m
Kronenbreite | crest width	7 m
Fundamentbreite | foundation width	24 m
Kronenkote | crest elevation	1957 m ü. M. | MASL
Stausee | reservoir lake	
Stauvolumen | reservoir volume	40'800'000 m³
Oberfläche | surface	812'000 m²

Davosersee	
Stauanlage | dam system	
Standort | location	Davos
Baujahr | construction year	- - - -
Sperrentyp | dam type	Natursee | natural lake
Bauvolumen | construction volume	- - - -
Sperrenhöhe | dam height	- - - -
Kronenlänge | crest length	- - - -
Kronenbreite | crest width	- - - -
Fundamentbreite | foundation width	- - - -
Kronenkote | crest elevation	1559 m ü. M. | MASL
Stausee | reservoir lake	
Stauvolumen | reservoir volume	11'500'000 m³
Oberfläche | surface	590'000 m²

Gebidem Stausee	
Stauanlage | dam system	
Standort | location	Blatten bei Naters
Baujahr | construction year	1964
Sperrentyp | dam type	Bogenmauer | arch dam
Bauvolumen | construction volume	228'000 m³
Sperrenhöhe | dam height	122 m
Kronenlänge | crest length	327 m
Kronenbreite | crest width	3 m
Fundamentbreite | foundation width	17 m
Kronenkote | crest elevation	1439 m ü. M. | MASL
Stausee | reservoir lake	
Stauvolumen | reservoir volume	9'200'000 m³
Oberfläche | surface	210'000 m²

Lac des Dix	
Stauanlage | dam system	
Standort | location	Hérémence
Baujahr | construction year	1961
Sperrentyp | dam type	Gewichtsmauer | gravity wall
Bauvolumen | construction volume	5'960'000 m³
Sperrenhöhe | dam height	285 m
Kronenlänge | crest length	695 m
Kronenbreite | crest width	15 m
Fundamentbreite | foundation width	200 m
Kronenkote | crest elevation	2365 m ü. M. | MASL
Stausee | reservoir lake	
Stauvolumen | reservoir volume	400'000'000 m³
Oberfläche | surface	4'040'000 m²

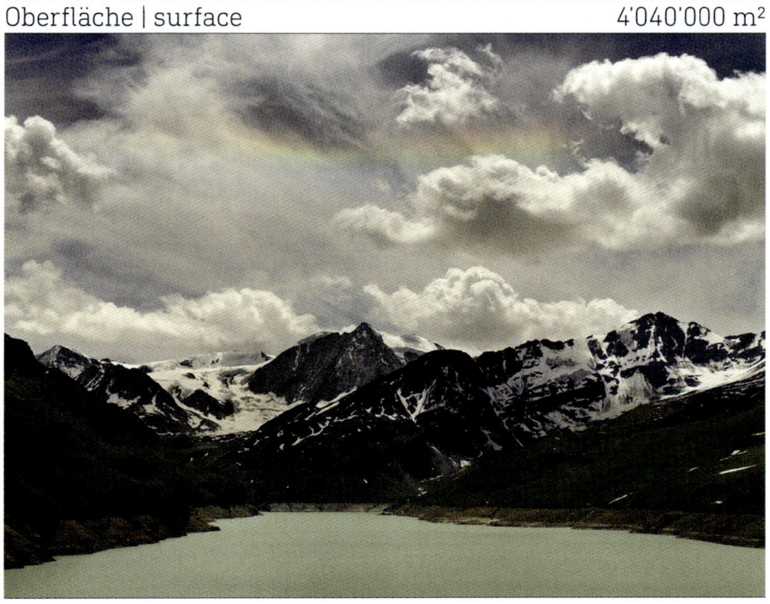

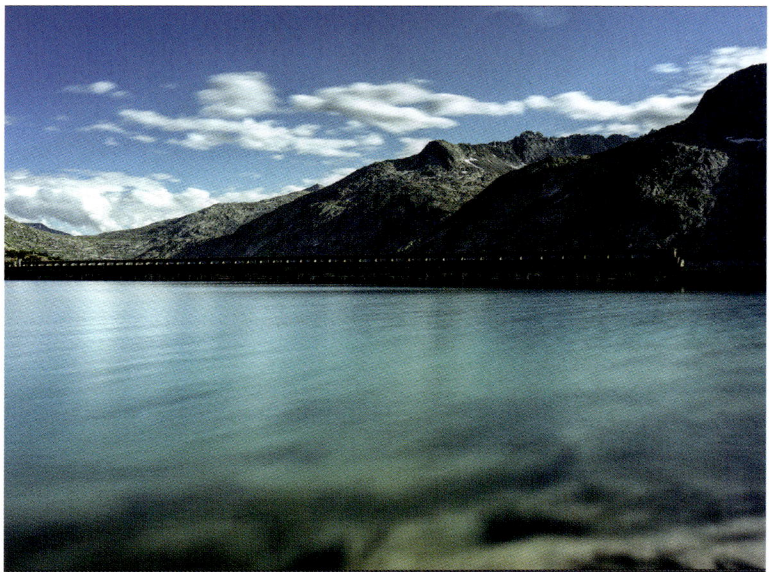

Gelmersee	
Stauanlage | dam system	
Standort | location	Guttannen
Baujahr | construction year	1929
Sperrentyp | dam type	Gewichtsmauer | gravity wall
Bauvolumen | construction volume	81'000 m³
Sperrenhöhe | dam height	35 m
Kronenlänge | crest length	370 m
Kronenbreite | crest width	3 m
Fundamentbreite | foundation width	22 m
Kronenkote | crest elevation	1850 m ü. M. | MASL
Stausee | reservoir lake	
Stauvolumen | reservoir volume	13'200'000 m³
Oberfläche | surface	645'000 m²

Lac d'Émosson	
Stauanlage | dam system	
Standort | location	Finhaut
Baujahr | construction year	1974
Sperrentyp | dam type	Bogenmauer | arch dam
Bauvolumen | construction volume	1'100'000 m³
Sperrenhöhe | dam height	180 m
Kronenlänge | crest length	554 m
Kronenbreite | crest width	9 m
Fundamentbreite | foundation width	48.5 m
Kronenkote | crest elevation	1930 m ü. M. | MASL
Stausee | reservoir lake	
Stauvolumen | reservoir volume	227'000'000 m³
Oberfläche | surface	3'270'000 m²

Gigerwaldsee	
Stauanlage | dam system	
Standort | location	Vättis
Baujahr | construction year	1976
Sperrentyp | dam type	Bogenmauer | arch dam
Bauvolumen | construction volume	446'000 m³
Sperrenhöhe | dam height	147 m
Kronenlänge | crest length	430 m
Kronenbreite | crest width	7 m
Fundamentbreite | foundation width	22 m
Kronenkote | crest elevation	1336 m ü. M. | MASL
Stausee | reservoir lake	
Stauvolumen | reservoir volume	33'400'000 m³
Oberfläche | surface	710'000 m²

Griessee	
Stauanlage | dam system	
Standort | location	Obergoms
Baujahr | construction year	1965
Sperrentyp | dam type	Schwergewichtsmauer | gravity dam
Bauvolumen | construction volume	259'000 m³
Sperrenhöhe | dam height	60 m
Kronenlänge | crest length	400 m
Kronenbreite | crest width	4.5 m
Fundamentbreite | foundation width	44 m
Kronenkote | crest elevation	2387 m ü. M. | MASL
Stausee | reservoir lake	
Stauvolumen | reservoir volume	18'700'000 m³
Oberfläche | surface	560'000 m²

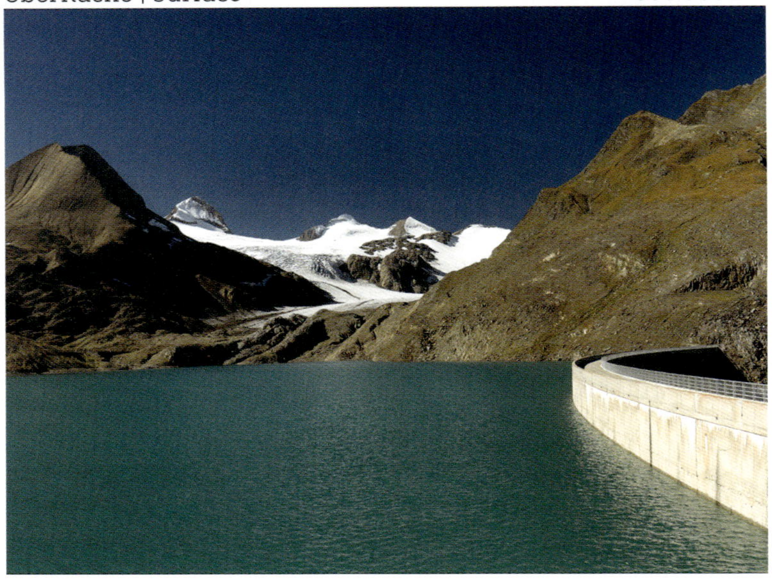

Grimselsee	
Stauanlage | dam system	
Standort | location	Guttannen
Baujahr | construction year	1932
Sperrentyp | dam type	Bogengewichtsmauer | arch-gravity dam
Bauvolumen | construction volume	340'000 m³
Sperrenhöhe | dam height	114 m
Kronenlänge | crest length	258 m
Kronenbreite | crest width	4 m
Fundamentbreite | foundation width	64 m
Kronenkote | crest elevation	1911 m ü. M. | MASL
Stausee | reservoir lake	
Stauvolumen | reservoir volume	100'490'000 m³
Oberfläche | surface	2'720'000 m²

Göscheneralpsee	
Stauanlage | dam system	
Standort | location	Göschenen
Baujahr | construction year	1960
Sperrentyp | dam type	Erdschüttdamm | earth fill dam
Bauvolumen | construction volume	9'300'000 m³
Sperrenhöhe | dam height	155 m
Kronenlänge | crest length	540 m
Kronenbreite | crest width	11 m
Fundamentbreite | foundation width	700 m
Kronenkote | crest elevation	1797 m ü. M. | MASL
Stausee | reservoir lake	
Stauvolumen | reservoir volume	76'000'000 m³
Oberfläche | surface	1'320'000 m²

Lac de l'Hongrin	
Stauanlage | dam system	
Standort | location	Ormont-Dessous
Baujahr | construction year	1971
Sperrentyp | dam type	Doppelte Bogenmauer | double arch dam
Bauvolumen | construction volume	345'000 m³
Sperrenhöhe | dam height	95 m, 123 m
Kronenlänge | crest length	600 m
Kronenbreite | crest width	3 m
Fundamentbreite | foundation width	18.4 m
Kronenkote | crest elevation	1255 m ü. M. | MASL
Stausee | reservoir lake	
Stauvolumen | reservoir volume	52'000'000 m³
Oberfläche | surface	1'600'000 m²

Klöntalersee	
Stauanlage | dam system	
Standort | location	Glarus
Baujahr | construction year	1908
Sperrentyp | dam type	Erddamm | earth dam
Bauvolumen | construction volume	110'000 m³
Sperrenhöhe | dam height	30 m
Kronenlänge | crest length	217 m
Kronenbreite | crest width	6 m
Fundamentbreite | foundation width	83 m
Kronenkote | crest elevation	850 m ü. M. | MASL
Stausee | reservoir lake	
Stauvolumen | reservoir volume	39'800'000 m³
Oberfläche | surface	3'300'000 m²

Lago di Lei	
Stauanlage | dam system	
Standort | location	Ferrera
Baujahr | construction year	1961
Sperrentyp | dam type	Bogenmauer | arch dam
Bauvolumen | construction volume	850'000 m³
Sperrenhöhe | dam height	138 m
Kronenlänge | crest length	635 m
Kronenbreite | crest width	15 m
Fundamentbreite | foundation width	28 m
Kronenkote | crest elevation	1932 m ü. M. | MASL
Stausee | reservoir lake	
Stauvolumen | reservoir volume	200'000'000 m³
Oberfläche | surface	4'120'000 m²

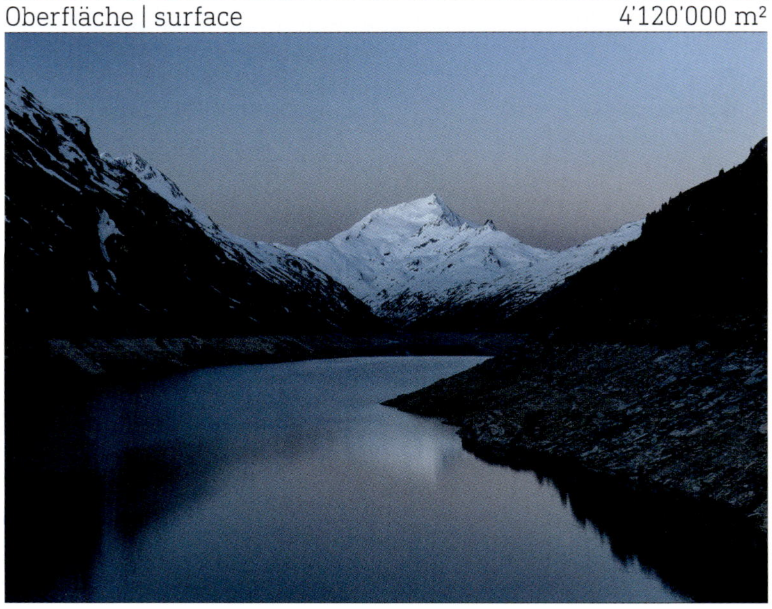

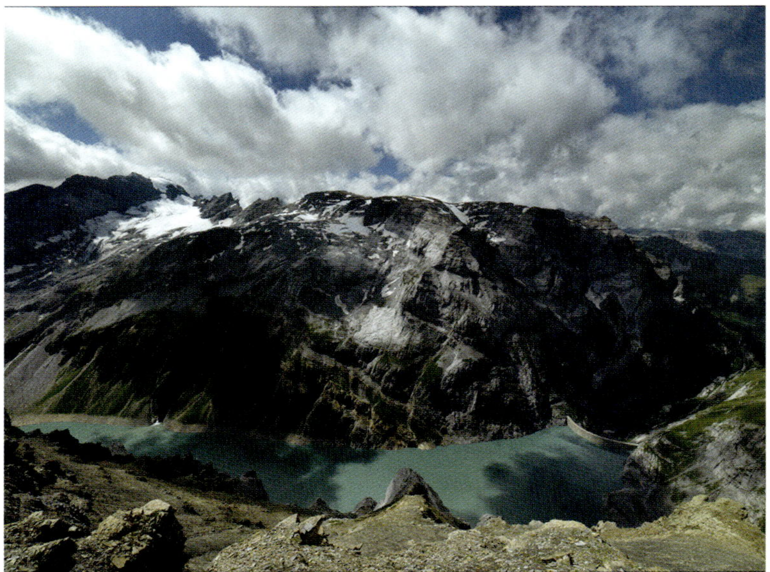

Limmerensee	
Stauanlage | dam system	
Standort | location	Glarus Süd
Baujahr | construction year	1962
Sperrentyp | dam type	Bogenmauer | arch dam
Bauvolumen | construction volume	553'000 m³
Sperrenhöhe | dam height	146 m
Kronenlänge | crest length	370 m
Kronenbreite | crest width	9 m
Fundamentbreite | foundation width	25 m
Kronenkote | crest elevation	1858 m ü. M. | MASL
Stausee | reservoir lake	
Stauvolumen | reservoir volume	92'000'000 m³
Oberfläche | surface	1'360'000 m²

Lago di Livigno	
Stauanlage | dam system	
Standort | location	Zernez
Baujahr | construction year	1968
Sperrentyp | dam type	Bogenmauer | arch dam
Bauvolumen | construction volume	776'000 m³
Sperrenhöhe | dam height	130 m
Kronenlänge | crest length	540 m
Kronenbreite | crest width	10–12 m
Fundamentbreite | foundation width	24.5 m
Kronenkote | crest elevation	1805 m ü. M. | MASL
Stausee | reservoir lake	
Stauvolumen | reservoir volume	164'600'000 m³
Oberfläche | surface	4'710'000 m²

Lago di Lucendro	
Stauanlage | dam system	
Standort | location	Airolo
Baujahr | construction year	1947
Sperrentyp | dam type	Pfeilermauer | buttressed wall
Bauvolumen | construction volume	154'000 m³
Sperrenhöhe | dam height	73 m
Kronenlänge | crest length	269 m
Kronenbreite | crest width	3.5 m
Fundamentbreite | foundation width	60 m
Kronenkote | crest elevation	2135 m ü. M. | MASL
Stausee | reservoir lake	
Stauvolumen | reservoir volume	25'000'000 m³
Oberfläche | surface	540'000 m²

Lago di Luzzone	
Stauanlage | dam system	
Standort | location	Blenio
Baujahr | construction year	1963
Sperrentyp | dam type	Bogenmauer | arch dam
Bauvolumen | construction volume	1'410'000 m³
Sperrenhöhe | dam height	225 m
Kronenlänge | crest length	510 m
Kronenbreite | crest width	4.6 m
Fundamentbreite | foundation width	36 m
Kronenkote | crest elevation	1609 m ü. M. | MASL
Stausee | reservoir lake	
Stauvolumen | reservoir volume	108'000'000 m³
Oberfläche | surface	1'270'000 m²

Bacino di Val Malvaglia	
Stauanlage | dam system	
Standort | location	Malvaglia
Baujahr | construction year	1959
Sperrentyp | dam type	Bogenmauer | arch dam
Bauvolumen | construction volume	162'000 m³
Sperrenhöhe | dam height	92 m
Kronenlänge | crest length	292 m
Kronenbreite | crest width	4 m
Fundamentbreite | foundation width	15.4 m
Kronenkote | crest elevation	992 m ü. M. | MASL
Stausee | reservoir lake	
Stauvolumen | reservoir volume	4'000'000 m³
Oberfläche | surface	189'000 m²

Lai da Marmorera	
Stauanlage | dam system	
Standort | location	Surses
Baujahr | construction year	1954
Sperrentyp | dam type	Erddamm | earth dam
Bauvolumen | construction volume	2'700'000 m³
Sperrenhöhe | dam height	91 m
Kronenlänge | crest length	400 m
Kronenbreite | crest width	12 m
Fundamentbreite | foundation width	400 m
Kronenkote | crest elevation	1684 m ü. M. | MASL
Stausee | reservoir lake	
Stauvolumen | reservoir volume	62'700'000 m³
Oberfläche | surface	1'410'000 m²

Mattmarksee	
Stauanlage | dam system	
Standort | location	Saas-Almagell
Baujahr | construction year	1969
Sperrentyp | dam type	Erdschüttdamm | earth fill dam
Bauvolumen | construction volume	10'500'000 m³
Sperrenhöhe | dam height	117 m
Kronenlänge | crest length	780 m
Kronenbreite | crest width	11 m
Fundamentbreite | foundation width	250–450 m
Kronenkote | crest elevation	2197 m ü. M. | MASL
Stausee | reservoir lake	
Stauvolumen | reservoir volume	100'000'000 m³
Oberfläche | surface	1'760'000 m²

Lac de Mauvoisin	
Stauanlage | dam system	
Standort | location	Bagnes
Baujahr | construction year	1957
Sperrentyp | dam type	Bogenmauer | arch dam
Bauvolumen | construction volume	2'030'000 m³
Sperrenhöhe | dam height	250 m
Kronenlänge | crest length	520 m
Kronenbreite | crest width	12 m
Fundamentbreite | foundation width	53.5 m
Kronenkote | crest elevation	1976 m ü. M. | MASL
Stausee | reservoir lake	
Stauvolumen | reservoir volume	210'000'000 m³
Oberfläche | surface	2'080'000 m²

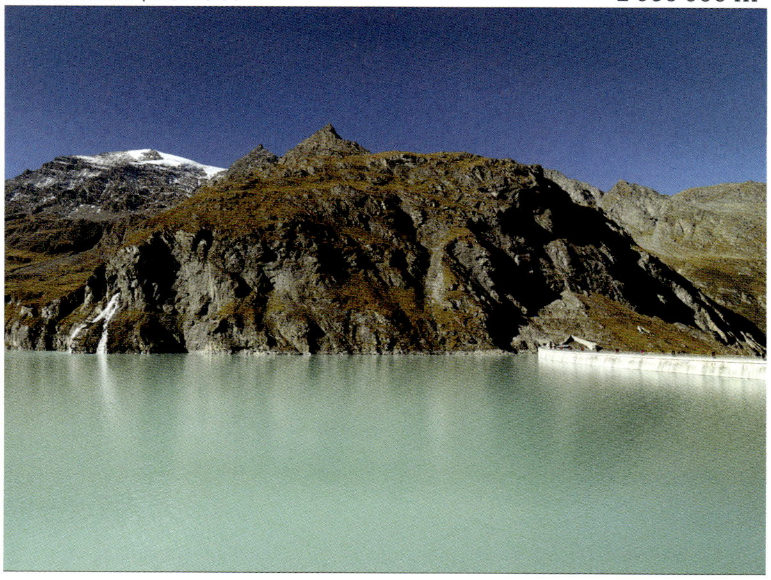

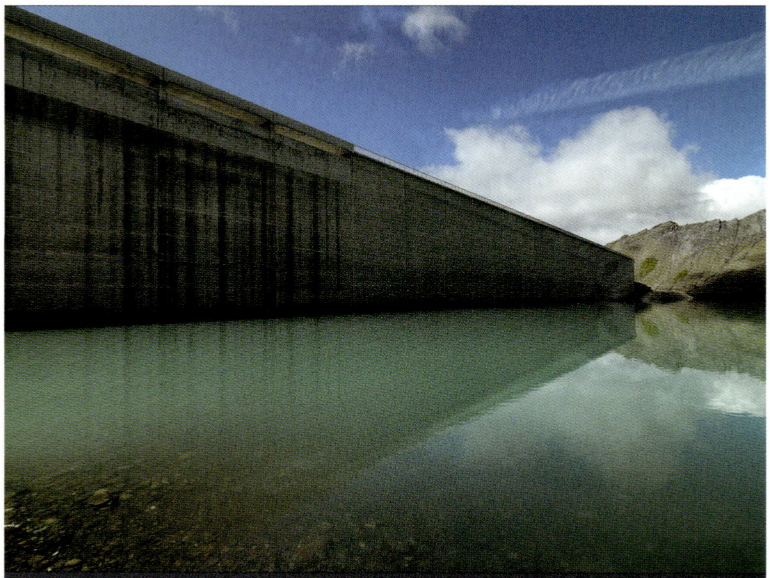

Muttsee	
Stauanlage | dam system	
Standort | location	Glarus Süd
Baujahr | construction year	2014
Sperrentyp | dam type	Schwergewichtsmauer | gravity dam
Bauvolumen | construction volume	250'000 m³
Sperrenhöhe | dam height	35 m
Kronenlänge | crest length	1'054 m
Kronenbreite | crest width	4–6 m
Fundamentbreite | foundation width	27 m
Kronenkote | crest elevation	2476 m ü. M. | MASL
Stausee | reservoir lake	
Stauvolumen | reservoir volume	23'000'000 m³
Oberfläche | surface	610'000 m²

Lac de Moiry	
Stauanlage | dam system	
Standort | location	Anniviers
Baujahr | construction year	1958
Sperrentyp | dam type	Bogenmauer | arch dam
Bauvolumen | construction volume	814'500 m³
Sperrenhöhe | dam height	148 m
Kronenlänge | crest length	610 m
Kronenbreite | crest width	7 m
Fundamentbreite | foundation width	34 m
Kronenkote | crest elevation	2250 m ü. M. | MASL
Stausee | reservoir lake	
Stauvolumen | reservoir volume	77'000'000 m³
Oberfläche | surface	1'300'000 m²

Lac de Montsalvens	
Stauanlage | dam system	
Standort | location	Broc, Charmey
Baujahr | construction year	1921
Sperrentyp | dam type	Bogenmauer | arch dam
Bauvolumen | construction volume	26'000 m³
Sperrenhöhe | dam height	52 m
Kronenlänge | crest length	110 m
Kronenbreite | crest width	2 m
Fundamentbreite | foundation width	22 m
Kronenkote | crest elevation	802 m ü. M. | MASL
Stausee | reservoir lake	
Stauvolumen | reservoir volume	12'600'000 m³
Oberfläche | surface	693'000 m²

Lai da Nalps	
Stauanlage | dam system	
Standort | location	Tujetsch
Baujahr | construction year	1962
Sperrentyp | dam type	Bogenmauer | arch dam
Bauvolumen | construction volume	593'000 m³
Sperrenhöhe | dam height	127 m
Kronenlänge | crest length	480 m
Kronenbreite | crest width	7 m
Fundamentbreite | foundation width	23 m
Kronenkote | crest elevation	1909 m ü. M. | MASL
Stausee | reservoir lake	
Stauvolumen | reservoir volume	44'500'000 m³
Oberfläche | surface	911'000 m²

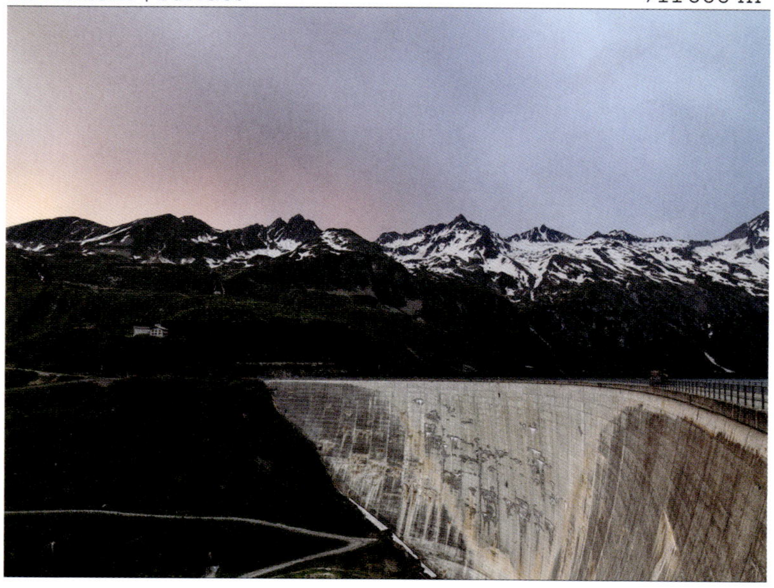

Lago del Narèt	
Stauanlage | dam system	
Standort | location	Lavizzara
Baujahr | construction year	1969
Sperrentyp | dam type	Bogenmauer | arch dam
Bauvolumen | construction volume	303'000 m³
Sperrenhöhe | dam height	80 m
Kronenlänge | crest length	435 m
Kronenbreite | crest width	7 m
Fundamentbreite | foundation width	17 m
Kronenkote | crest elevation	2311 m ü. M. | MASL
Stausee | reservoir lake	
Stauvolumen | reservoir volume	31'600'000 m³
Oberfläche | surface	860'000 m²

Lac de Pérolles	
Stauanlage | dam system	
Standort | location	Fribourg
Baujahr | construction year	1872
Sperrentyp | dam type	Gewichtsmauer | gravity wall
Bauvolumen | construction volume	34'000 m³
Sperrenhöhe | dam height	24 m
Kronenlänge | crest length	113.5 m
Kronenbreite | crest width	4.5 m
Fundamentbreite | foundation width	23 m
Kronenkote | crest elevation	553 m ü. M. | MASL
Stausee | reservoir lake	
Stauvolumen | reservoir volume	600'000 m³
Oberfläche | surface	350'000 m²

Oberaarsee	
Stauanlage | dam system	
Standort | location	Guttannen
Baujahr | construction year	1953
Sperrentyp | dam type	Gewichtsmauer | gravity wall
Bauvolumen | construction volume	453'000 m³
Sperrenhöhe | dam height	104 m
Kronenlänge | crest length	526 m
Kronenbreite | crest width	4.4 m
Fundamentbreite | foundation width	75 m
Kronenkote | crest elevation	2304 m ü. M. | MASL
Stausee | reservoir lake	
Stauvolumen | reservoir volume	13'000'000 m³
Oberfläche | surface	1'470'000 m²

Räterichsbodensee	
Stauanlage | dam system	
Standort | location	Guttannen
Baujahr | construction year	1950
Sperrentyp | dam type	Gewichtsmauer | gravity wall
Bauvolumen | construction volume	230'000 m³
Sperrenhöhe | dam height	92 m
Kronenlänge | crest length	456 m
Kronenbreite | crest width	4 m
Fundamentbreite | foundation width	78 m
Kronenkote | crest elevation	1768 m ü. M. | MASL
Stausee | reservoir lake	
Stauvolumen | reservoir volume	26'300'000 m³
Oberfläche | surface	670'000 m²

Lai da Ova Spin	
Stauanlage | dam system	
Standort | location	Zernez
Baujahr | construction year	1968
Sperrentyp | dam type	Bogenmauer | arch dam
Bauvolumen | construction volume	27'000 m³
Sperrenhöhe | dam height	73 m
Kronenlänge | crest length	130 m
Kronenbreite | crest width	3 m
Fundamentbreite | foundation width	8 m
Kronenkote | crest elevation	1633 m ü. M. | MASL
Stausee | reservoir lake	
Stauvolumen | reservoir volume	6'240'000 m³
Oberfläche | surface	360'000 m²

Panixersee	
Stauanlage | dam system	
Standort | location	Ilanz/Glion
Baujahr | construction year	1900
Sperrentyp | dam type	Gewichtsmauer | gravity wall
Bauvolumen | construction volume	161'000 m³
Sperrenhöhe | dam height	53 m
Kronenlänge | crest length	270 m
Kronenbreite | crest width	5 m
Fundamentbreite | foundation width	55.7 m
Kronenkote | crest elevation	1450 m ü. M. | MASL
Stausee | reservoir lake	
Stauvolumen | reservoir volume	7'300'000 m³
Oberfläche | surface	400'000 m²

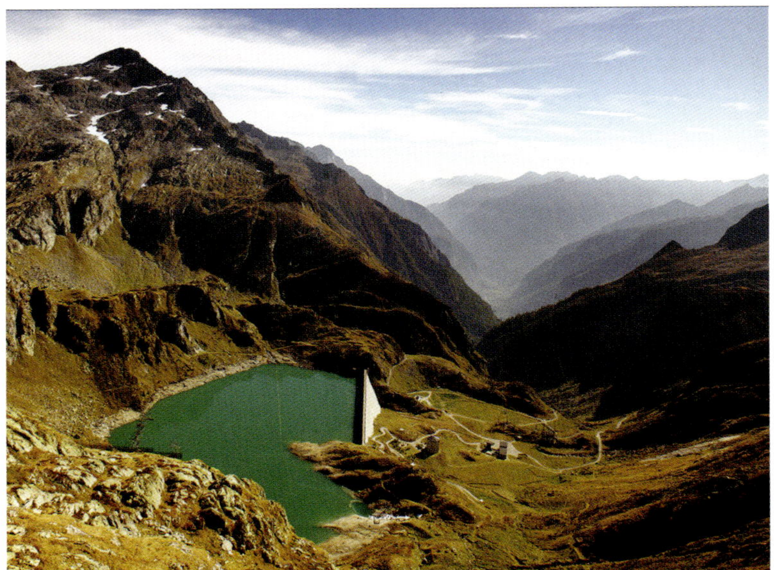

Lago di Robièi	
Stauanlage | dam system	
Standort | location	Cevio
Baujahr | construction year	1967
Sperrentyp | dam type	Gewichtsmauer | gravity wall
Bauvolumen | construction volume	180'000 m³
Sperrenhöhe | dam height	68 m
Kronenlänge | crest length	355 m
Kronenbreite | crest width	4.6 m
Fundamentbreite | foundation width	50 m
Kronenkote | crest elevation	1941 m ü. M. | MASL
Stausee | reservoir lake	
Stauvolumen | reservoir volume	6'900'000 m³
Oberfläche | surface	240'000 m²

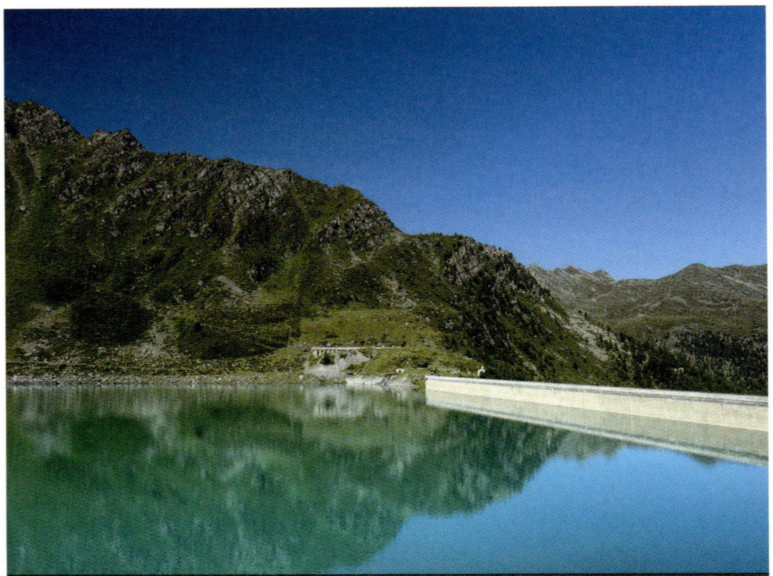

Lac de Salanfe	
Stauanlage | dam system	
Standort | location	Evionnaz
Baujahr | construction year	1947
Sperrentyp | dam type	Gewichtsmauer | gravity wall
Bauvolumen | construction volume	230'000 m³
Sperrenhöhe | dam height	52 m
Kronenlänge | crest length	616 m
Kronenbreite | crest width	5 m
Fundamentbreite | foundation width	40 m
Kronenkote | crest elevation	1925 m ü. M. | MASL
Stausee | reservoir lake	
Stauvolumen | reservoir volume	40'000'000 m³
Oberfläche | surface	1'850'000 m²

Lago del Sambuco	
Stauanlage | dam system	
Standort | location	Lavizzara
Baujahr | construction year	1956
Sperrentyp | dam type	Bogenmauer | arch dam
Bauvolumen | construction volume	775'000 m³
Sperrenhöhe | dam height	130 m
Kronenlänge | crest length	363 m
Kronenbreite | crest width	8 m
Fundamentbreite | foundation width	71 m
Kronenkote | crest elevation	1462 m ü. M. | MASL
Stausee | reservoir lake	
Stauvolumen | reservoir volume	64'000'000 m³
Oberfläche | surface	1'110'000 m²

Lago della Sella	
Stauanlage | dam system	
Standort | location	Airolo
Baujahr | construction year	1947
Sperrentyp | dam type	Gewichtsmauer | gravity wall
Bauvolumen | construction volume	75'000 m³
Sperrenhöhe | dam height	36 m
Kronenlänge | crest length	334 m
Kronenbreite | crest width	3.5 m
Fundamentbreite | foundation width	25 m
Kronenkote | crest elevation	2256 m ü. M. | MASL
Stausee | reservoir lake	
Stauvolumen | reservoir volume	9'200'000 m³
Oberfläche | surface	450'000 m²

Schiffenensee	
Stauanlage | dam system	
Standort | location	Düdingen
Baujahr | construction year	1964
Sperrentyp | dam type	Bogenmauer | arch dam
Bauvolumen | construction volume	185'000 m³
Sperrenhöhe | dam height	47 m
Kronenlänge | crest length	417 m
Kronenbreite | crest width	7 m
Fundamentbreite | foundation width	12 m
Kronenkote | crest elevation	534 m ü. M. | MASL
Stausee | reservoir lake	
Stauvolumen | reservoir volume	58'600'000 m³
Oberfläche | surface	4'250'000 m²

Lai da Sontga Maria	
Stauanlage | dam system	
Standort | location	Medel
Baujahr | construction year	1968
Sperrentyp | dam type	Bogenmauer | arch dam
Bauvolumen | construction volume	654'000 m³
Sperrenhöhe | dam height	117 m
Kronenlänge | crest length	560 m
Kronenbreite | crest width	8 m
Fundamentbreite | foundation width	21 m
Kronenkote | crest elevation	1910 m ü. M. | MASL
Stausee | reservoir lake	
Stauvolumen | reservoir volume	67'000'000 m³
Oberfläche | surface	1'774'000 m²

Sufnersee	
Stauanlage | dam system	
Standort | location	Sufers
Baujahr | construction year	1962
Sperrentyp | dam type	Bogenmauer | arch dam
Bauvolumen | construction volume	22'100 m³
Sperrenhöhe | dam height	58 m
Kronenlänge | crest length	125 m
Kronenbreite | crest width	3 m
Fundamentbreite | foundation width	7.9 m
Kronenkote | crest elevation	1403 m ü. M. | MASL
Stausee | reservoir lake	
Stauvolumen | reservoir volume	21'400'000 m³
Oberfläche | surface	940'000 m²

Lac de Tseuzier	
Stauanlage | dam system	
Standort | location	Ayent, Icogne VS
Baujahr | construction year	1957
Sperrentyp | dam type	Bogenmauer | arch dam
Bauvolumen | construction volume	300'000 m³
Sperrenhöhe | dam height	156 m
Kronenlänge | crest length	256 m
Kronenbreite | crest width	- - - -
Fundamentbreite | foundation width	- - - -
Kronenkote | crest elevation	1779 m ü. M. | MASL
Stausee | reservoir lake	
Stauvolumen | reservoir volume	126'000'000 m³
Oberfläche | surface	850'000 m²

Turtmannsee	
Stauanlage | dam system	
Standort | location	Oberems, Turtmann-Unterems
Baujahr | construction year	1958
Sperrentyp | dam type	Bogenmauer | arch dam
Bauvolumen | construction volume	3'200 m³
Sperrenhöhe | dam height	33 m
Kronenlänge | crest length	115 m
Kronenbreite | crest width	1.2 m
Fundamentbreite | foundation width	7 m
Kronenkote | crest elevation	2177 m ü. M. | MASL
Stausee | reservoir lake	
Stauvolumen | reservoir volume	780'000 m³
Oberfläche | surface	1'200'000 m²

Vordersee	
Stauanlage | dam system	
Standort | location	Fieschertal
Baujahr | construction year	1986
Sperrentyp | dam type	Erddamm | earth dam
Bauvolumen | construction volume	48'850 m³
Sperrenhöhe | dam height	16 m
Kronenlänge | crest length	174 m
Kronenbreite | crest width	4 m
Fundamentbreite | foundation width	60 m
Kronenkote | crest elevation	2363 m ü. M. | MASL
Stausee | reservoir lake	
Stauvolumen | reservoir volume	550'000 m³
Oberfläche | surface	3'200 m²

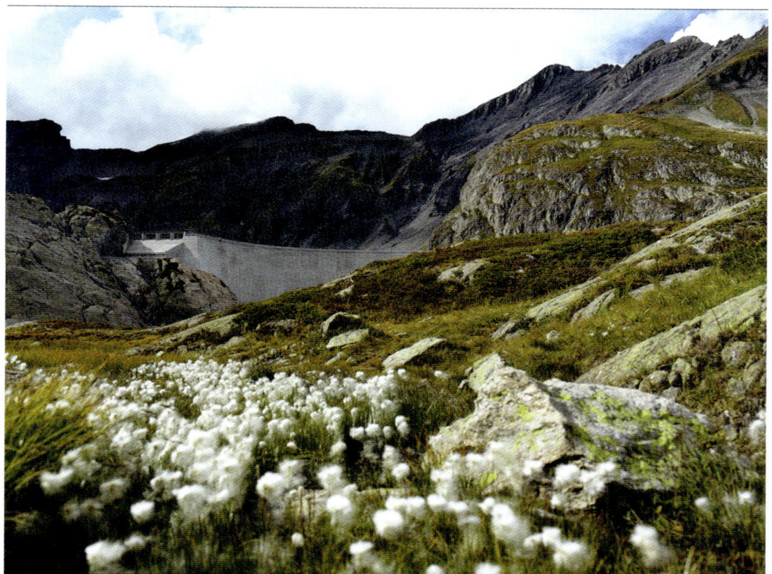

Lac du Vieux Emosson	
Stauanlage | dam system	
Standort | location	Finhaut
Baujahr | construction year	1955/2014
Sperrentyp | dam type	Bogenmauer | arch dam
Bauvolumen | construction volume	62'000 m³
Sperrenhöhe | dam height	76.5 m
Kronenlänge | crest length	205 m
Kronenbreite | crest width	4.2–4.5 m
Fundamentbreite | foundation width	17.75 m
Kronenkote | crest elevation	2226.5 m ü. M. | MASL
Stausee | reservoir lake	
Stauvolumen | reservoir volume	25'000'000 m³
Oberfläche | surface	550'000 m²

Wägitalersee	
Stauanlage | dam system	
Standort | location	Innerthal
Baujahr | construction year	1924
Sperrentyp | dam type	Gewichtsmauer | gravity wall
Bauvolumen | construction volume	237'000 m³
Sperrenhöhe | dam height	110 m
Kronenlänge | crest length	156 m
Kronenbreite | crest width	4 m
Fundamentbreite | foundation width	75 m
Kronenkote | crest elevation	902 m ü. M. | MASL
Stausee | reservoir lake	
Stauvolumen | reservoir volume	150'000'000 m³
Oberfläche | surface	4'180'000 m²

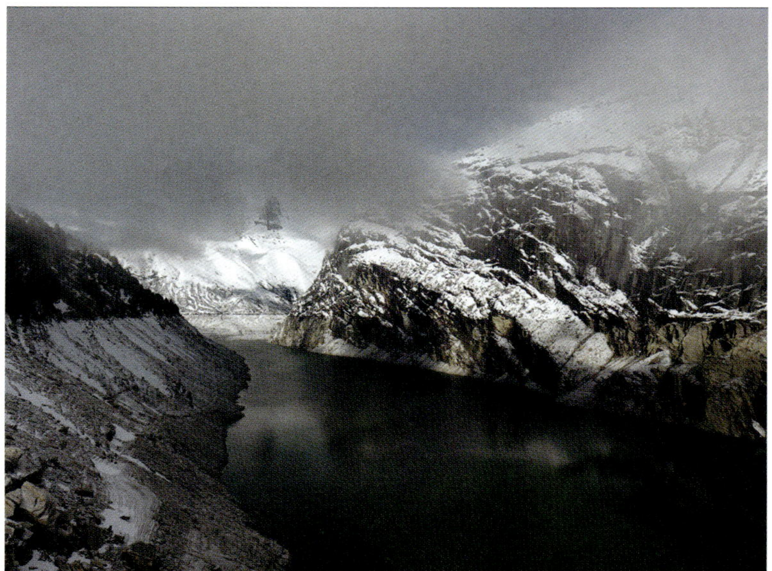

Zervreilasee	
Stauanlage | dam system	
Standort | location	Vals
Baujahr | construction year	1957
Sperrentyp | dam type	Bogenmauer | arch dam
Bauvolumen | construction volume	650'000 m³
Sperrenhöhe | dam height	151 m
Kronenlänge | crest length	504 m
Kronenbreite | crest width	7 m
Fundamentbreite | foundation width	35 m
Kronenkote | crest elevation	1863 m ü. M. | MASL
Stausee | reservoir lake	
Stauvolumen | reservoir volume	100'000'000 m³
Oberfläche | surface	1'610'000 m²

Lago di Vogorno	
Stauanlage | dam system	
Standort | location	Gordola
Baujahr | construction year	1965
Sperrentyp | dam type	Bogenmauer | arch dam
Bauvolumen | construction volume	670'000 m³
Sperrenhöhe | dam height	220 m
Kronenlänge | crest length	380 m
Kronenbreite | crest width	7 m
Fundamentbreite | foundation width	25 m
Kronenkote | crest elevation	473.5 m ü. M. | MASL
Stausee | reservoir lake	
Stauvolumen | reservoir volume	105'000'000 m³
Oberfläche | surface	1'680'000 m²

Zu guter Letzt herzlichen Dank an alle, die «StauWerke» mit ihrem Geld, ihrer Arbeit sowie mit Worten und Gedanken unterstützt haben. Ein Langzeitprojekt wie dieses ist an einem Tag und von einem Kopf und zwei Händen allein nicht zu schaffen, weder draussen in der Natur noch drinnen im Atelier. Mein besonderer Dank geht an | Last but not least, I would like to sincerely thank everyone who supported "StauWerke" with their money, their efforts, words and thoughts. A long-term project like this one cannot be accomplished in a single day by a single head and two hands alone, neither outdoors in nature nor in the studio. My special thanks go out to

Evi Walther, Debora Walther, Pascal Walther, Ursina und Michael Walther sowie Commune d'Ayent (ayent.ch), Urs Bichler, Anny und Martin Bösch, Comune di Bregaglia (comunedibregaglia.ch), Eliane und Hans Eberhard, ewz – Elektrizitätswerk der Stadt Zürich (ewz.ch), Arlette Frias, Köbi Gantenbein (hochparterre.ch), Robert Geiter (geiterkonzept.ch), Reymond und Pilar Gmür, Ernst Göhner Stiftung (ernst-goehner-stiftung.ch), Gemeinde Göschenen (goeschenen.ch), Peter Hasler, Rita und Urs Kägi Feller, Urs Kühne (kuehne-treuhand.ch), Felice Iovino (iovino-ci.ch), Markus Mäder (nightwriter.ch), Philipp Metzger (extrawürste.ch), Mikro + Repro AG (mikrorepro.ch), Marc Pfyffer (pfyfferschmid.ch), Comune di Poschiavo (poschiavo.ch), Wenzel Ries, Erika und Hans Jörg Schaer, Schweizerischer Wasserwirtschaftsverband (swv.ch), Elmar Steiger (steiger-partner.ch), Damian Steiger, VUE – Verein für umweltgerechte Energie (naturemade.ch), Heidi und Herbert Walther, Michael Walther (michaelwalther.ch), Tom Werner, Markus Züger (zueger-beratung.ch) und Jeannine Zubler (jeanninezubler.com).

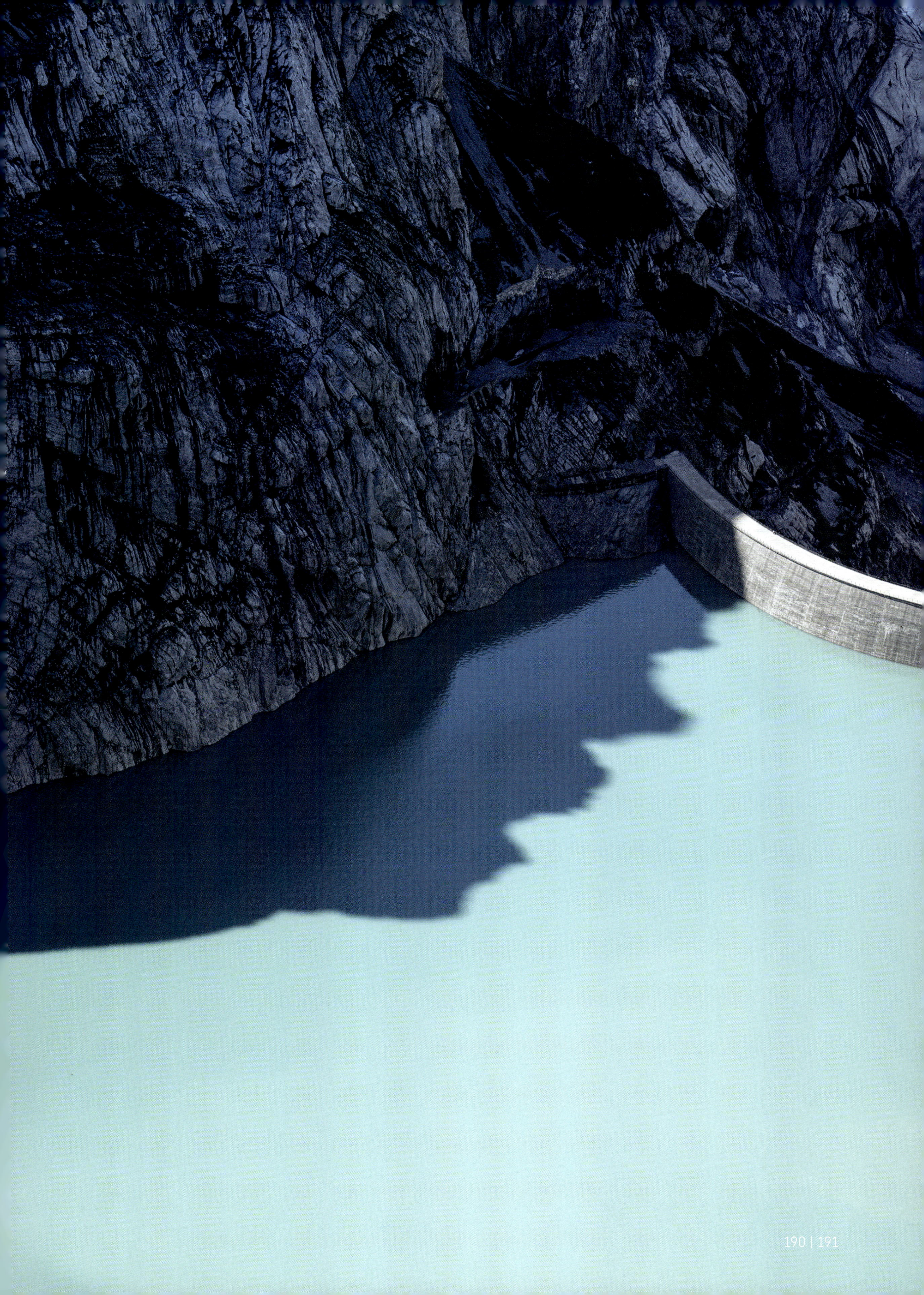

Umschlag	cover	Zervreilasee
Seite 3	page 3	Lago di Luzzone
Seite 4	page 4	Griessee
Seite 7	page 7	Gigerwaldsee
Seite 8	page 8	Limmerensee
Seite 10	page 10	Lago di Luzzone
Seite 12	page 12	Lägh da l'Albigna
Seite 14	page 14	Lac de Moiry
Seite 16	page 16	Bacino di Val Malvaglia
Seite 18	page 18	Lago dei Cavagnöö
Seite 20	page 20	Göscheneralpsee
Seite 23	page 23	Lac de Cleuson
Seite 24	page 24	Muttsee
Seite 26	page 26	Lai da Curnera
Seite 29	page 29	Lac de Moiry
Seite 30	page 30	Lai da Sontga Maria
Seite 33	page 33	Lai da Marmorera
Seite 36	page 36	Lai da Marmorera
Seite 40	page 40	Klöntalersee
Seite 41	page 41	Klöntalersee
Seite 42	page 42	Lago Bianco (Nord)
Seite 44	page 44	Lago di Lei
Seite 46	page 46	Oberaarsee
Seite 48	page 48	Oberaarsee
Seite 50	page 50	Limmerensee
Seite 52	page 52	Lac de Tseuzier
Seite 54	page 54	Lago di Luzzone
Seite 56	page 56	Muttsee
Seite 58	page 58	Lac de Tseuzier
Seite 60	page 60	Lago di Livigno
Seite 62	page 62	Lägh da l'Albigna
Seite 63	page 63	Lago Bianco (Nord)
Seite 64	page 64	Grimselsee
Seite 65	page 65	Lago del Sambuco
Seite 66	page 66	Turtmannsee
Seite 67	page 67	Lago di Robièi
Seite 68	page 68	Oberaarsee
Seite 70	page 70	Lägh da l'Albigna
Seite 72	page 72	Grimselsee
Seite 73	page 73	Grimselsee
Seite 74	page 74	Grimselsee
Seite 76	page 76	Lägh da l'Albigna
Seite 78	page 78	Lago di Luzzone
Seite 80	page 80	Grimselsee
Seite 82	page 82	Lac de l'Hongrin
Seite 83	page 83	Lac de l'Hongrin
Seite 84	page 84	Bacino di Val Malvaglia
Seite 86	page 86	Lai da Curnera
Seite 88	page 88	Vordersee
Seite 90	page 90	Lago del Sambuco
Seite 92	page 92	Davosersee
Seite 94	page 94	Wägitalersee
Seite 96	page 96	Lai da Ova Spin
Seite 98	page 98	Lac de Mauvoisin
Seite 100	page 100	Lac de Moiry
Seite 102	page 102	Schiffenensee
Seite 104	page 104	Grimselsee
Seite 105	page 105	Lago dei Cavagnöö
Seite 106	page 106	Lac de l'Hongrin
Seite 107	page 107	Lac de l'Hongrin
Seite 108	page 108	Lago Bianco (Nord)
Seite 110	page 110	Lago di Lucendro
Seite 111	page 111	Lago di Lucendro
Seite 112	page 112	Lac de Mauvoisin
Seite 113	page 113	Lac de Mauvoisin
Seite 114	page 114	Gebidem Stausee
Seite 115	page 115	Sufnersee
Seite 116	page 116	Lac de Salanfe
Seite 118	page 118	Lac du Vieux Emosson
Seite 120	page 120	Lac de Pérolles
Seite 121	page 121	Lago del Narèt
Seite 122	page 122	Limmerensee
Seite 124	page 124	Lac de Montsalvens
Seite 126	page 126	Lago di Vogorno
Seite 128	page 128	Lago di Vogorno
Seite 130	page 130	Lago di Robièi
Seite 131	page 131	Lago della Sella
Seite 132	page 132	Gigerwaldsee
Seite 133	page 133	Lac des Dix
Seite 134	page 134	Muttsee
Seite 136	page 136	Lac d'Émosson
Seite 138	page 138	Lac des Dix
Seite 140	page 140	Lac des Dix
Seite 142	page 142	Mattmarksee
Seite 143	page 143	Mattmarksee
Seite 144	page 144	Lac des Dix
Seite 145	page 145	Lac des Dix
Seite 146	page 146	Muttsee
Seite 147	page 147	Muttsee
Seite 149	page 149	Schiffenensee
Seite 150	page 150	Lac du Vieux Emosson
Seite 152	page 152	Lägh da l'Albigna
Seite 154	page 154	Lago del Narèt
Seite 156	page 156	Lago Bianco (Nord)
Seite 158	page 158	Räterichsbodensee
Seite 159	page 159	Räterichsbodensee
Seite 160	page 160	Gigerwaldsee
Seite 162	page 162	Panixersee
Seite 163	page 163	Panixersee
Seite 164	page 164	Lago del Narèt
Seite 166	page 166	Lago di Luzzone
Seite 168	page 168	Lago di Lei
Seite 170	page 170	Lai da Nalps
Seite 172	page 172	Lago dei Cavagnöö
Seite 174	page 174	Lago del Narèt
Seite 176	page 176	Gelmersee
Seite 178	page 178	Lac des Dix
Seite 179	page 179	Lac de Salanfe
Seite 181	page 181	Griessee
Seite 191	page 191	Limmerensee

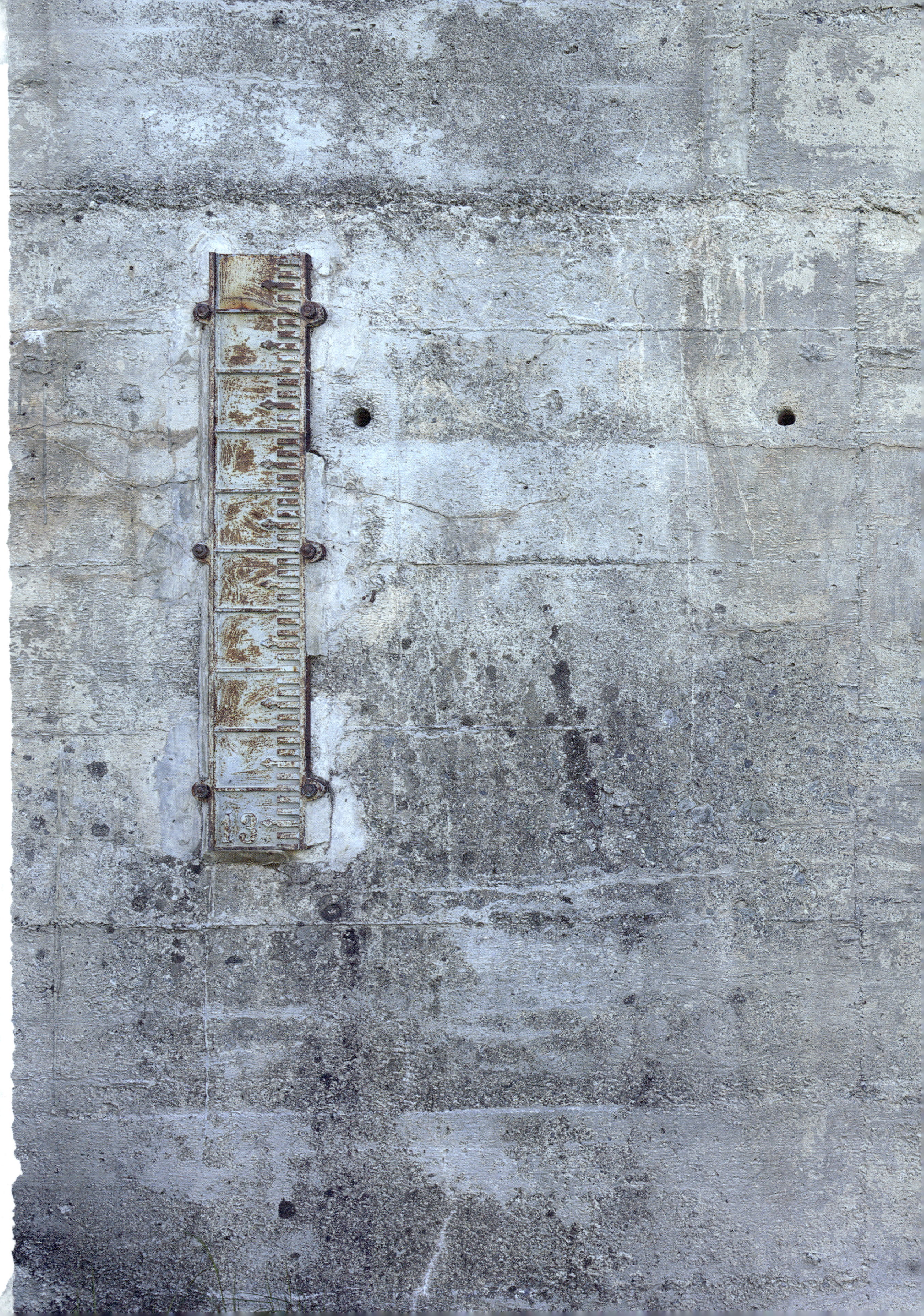
13

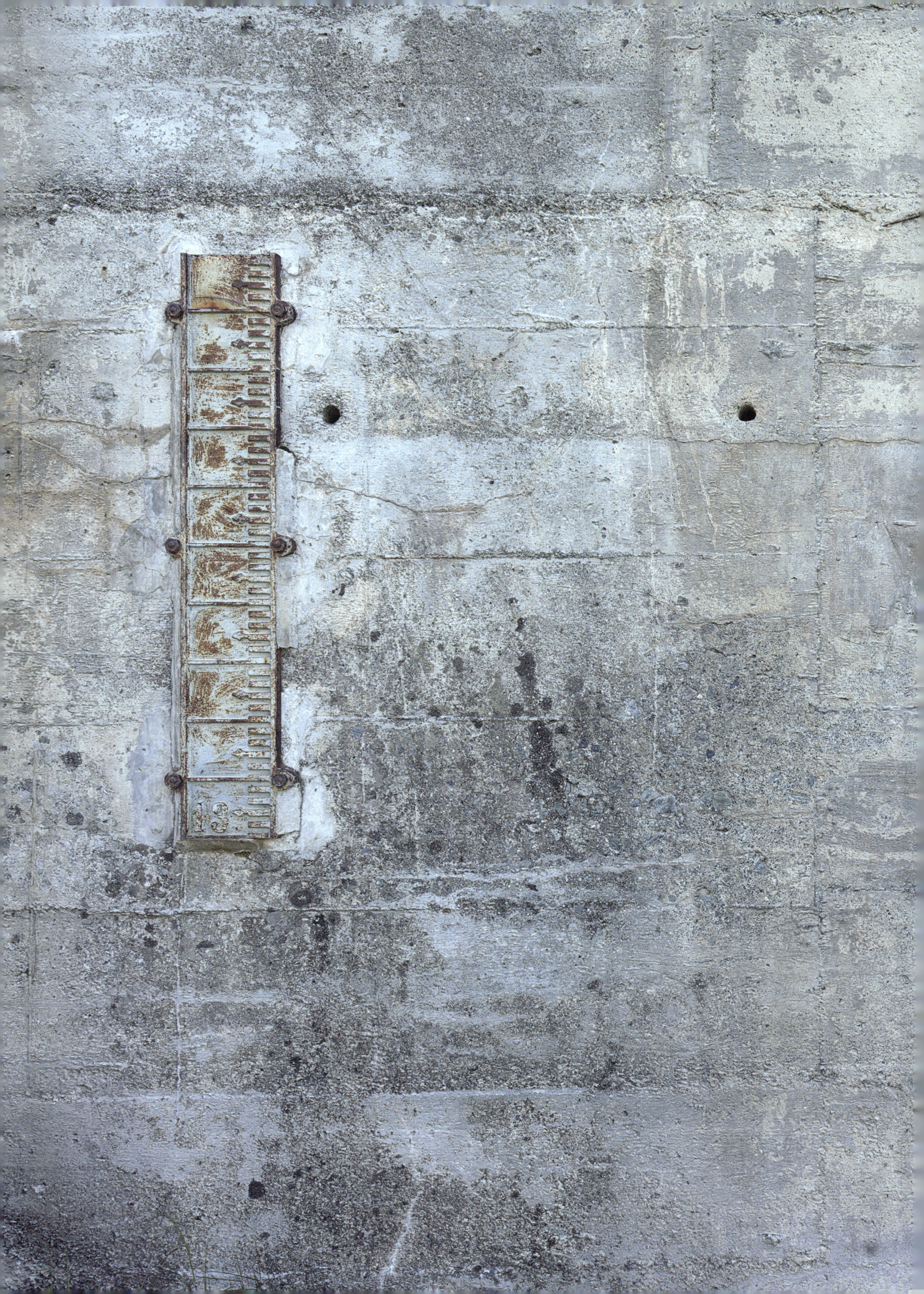
13